THE MONEY

"Inspirational. Living with [...] miserable; rather it left Ma [...] and, ultimately, wiser."

Rob Hopkins, author of *The Transition Handbook*
and founder of the Transition Movement

"It's difficult not to admire the philosophy and the infectious home-spun and passionate tone of the book."

Benedict Allen, TV survivalist and author of
The Faber Book of Exploration

"An inspiring and entertaining guide to escaping the money trap and reconnecting with reality."

Paul Kingsnorth, author of *Real England:
The Battle Against the Bland*

"Intriguing. It makes several very important points. A powerful look not so much into the past as into the future."

Bill McKibben, author of *Deep Economy*

"Thought-provoking, inspiring, and eye-opening. Mark writes from the heart with compelling honesty and humour. Fascinating."

Brigit Strawbridge, star of BBC's
It's Not Easy Being Green

"A pleasure to read. An important experiment which has relevance for everything from global sustainability to local community. Boyle tells it with wit, good humour, and generosity of spirit."

Patrick Whitefield, author of *The Earth Care Manual*

Mark Boyle is the author of *The Moneyless Man*, *The Moneyless Manifesto* and *Drinking Molotov Cocktails with Gandhi*, which have been translated into over twenty languages. His new book, *The Way Home*, is also published by Oneworld. A former business graduate, he lived entirely without money for three years. He has written columns for the *Guardian* and has irregularly contributed to international press, radio and television. He lives on a smallholding in Co. Galway, Ireland.

THE MONEY-LESS MAN

MARK BOYLE

ONEWORLD

A Oneworld Book

First published by Oneworld Publications 2010
This edition published by Oneworld Publications 2019
Reprinted, 2020

ISBN 978-1-78607-599-4
eISBN 978-1-85168-878-4

Typeset by Jayvee, Trivandrum, India
Printed and bound in Great Britain by Clays Ltd, Elcograf S.p.A.

Oneworld Publications
10 Bloomsbury Street
London, WC1B 3SR

Stay up to date with the latest books,
special offers, and exclusive content from
Oneworld with our newsletter

Sign up on our website
oneworld-publications.com

MIX
Paper from
responsible sources
FSC® C018072

To MKG

CONTENTS

CONTENTS

ACKNOWLEDGEMENTS

My name is on the front of this book, which suggests that these words are mine. But that's a half-truth. I claim no ownership. How could I? They are merely an accumulation of all that has come before them – the people I've met, the books I've read, the songs I grew up with, the rivers I've swum in, the girls I've kissed, the films I've watched, the traditions I've learned, the philosophers I've studied, the mistakes I've made, the violence I've seen and the love I've witnessed.

There are a few people really close to me to whom I'd really like to express my gratitude (disclaimer – if you are not named it doesn't mean I don't love you). My folks Marian and Josie, for giving me everything they ever could and for their relentless support. People like Chris and Suzie Adams (and wee Oak), Dawn, Markus and Olivia (to name just a few) who have helped me forge this path and who were there for me when I first walked down it, when I stumbled on it and who are still helping me today. To Mari, for your love and the unbreakable bond I have with you. To Fergus, for being a beacon of light in darkness and for reminding me why I do it. For those like Marty, Stephen and Gerard, who have taken different paths but who define the word 'friend' for me. To my community, near and far, whose wealth of knowledge, skills and friendship have had a value far beyond money over the last year. To Mike at Oneworld, my fantastic editor whom you need to thank if, for some strange reason, you end up enjoying this book and to Sallyanne for being the most supportive agent in the world.

Last, to the many thousands who have contacted me to offer their support over the year and to those who have criticised me, as it reminds me that my opinion is just one of many and that I have much to learn.

PROLOGUE

BUY NOTHING EVE, 28TH NOVEMBER 2008

The timing just doesn't get any better. Five past six on the evening of my last day in moneydom and as far as I am concerned, the shops have effectively closed down for a year. It's been an unexpectedly long day; the media caught a sniff of my plans to live without money and so, instead of making the final preparations for my impending social experiment and – far more important – having a last pint of ale at my local, I ended up doing interview after interview after interview. The sound of my voice answering the same questions over and over again has made me slightly nauseous.

Cycling home from my final interview at the BBC, on a short-cut through a particularly boozy, neon-lit and broken-glass-clad quarter of Bristol, I feel my rear end wobble. It's nothing major, only a puncture, but it is a symbolic example of the challenges I'll be facing every day for the next twelve months. I'm eighteen miles from my caravan, where I've foolishly left my repair kit, but I can stop at my girlfriend Claire's house to patch up the tube. My only concern is that I'll have to drag my slightly crocked bicycle, with two heavy panniers on the back, for just over three miles. Given that I'm five minutes too late to buy a new wheel, I could really do without buckling the one I have.

On my way there, I give my mate, Fergus Drennan, a shout. Fergus is a famous forager but, unfortunately, a terrible bike mechanic. Nevertheless, he's irrepressibly enthusiastic and just what I need. The pressure of time, mixed with apprehension about the year ahead, is starting to take its toll. After we eventually make it to Claire's house, as I mindlessly start undoing what I think is the back wheel, he describes how I can make paper and ink from mushrooms. Exhausted, but intrigued by his ramblings, I'm increasingly frustrated at my difficulty in getting the wheel off. At the precise moment I think I should get some food inside me before I either pass out or shove a Death Cap mushroom down Fergus's throat, there's a huge ping! and something that looks rather important springs across the room. Instead of loosening the wheel, in my exhaustion I've released the rear dérailleur. This really isn't very good news. With the exception of my body, this bike is easily the most important possession for my impending experiment. Actually it's not merely important, it's absolutely essential. It's a thirty-six-mile round trip on foot to many of my sources of potential food and wood and eighteen miles to most of my friends; without the bike, travelling to meetings would become impossible and I wouldn't have a hope in hell of being able to scavenge for the bits and pieces I'll inevitably need throughout the year.

I know a bit about bikes but something as intricate as the rear dérailleur is beyond me. In my previous, moneyed, existence, if something went seriously wrong with the bike, I took it to the bike shop, bought some new parts and paid the nice assistant to fix it. That, however, was no longer an option. I'd spent the day talking to reporters about how, for six months, I'd been preparing myself to succeed in living without money for a year and here I was, four hours before I'd officially started, lying, completely mentally and physically exhausted, beside a freshly-mangled bicycle that was at the heart of my plans. Given the fact that I was also due to cook a free three-course meal the next day for one hundred and fifty people, made from wild and urban-foraged foods that I hadn't yet gathered, I was starting to feel the strain.

It wasn't just the bicycle that worried me. It was one small example of the thousands of problems I encountered in a normal year. The difference was that in the past, I could have thrown money at my problems whenever and wherever they arose. I realised what a precarious position I was in, about to enter a world of which I had very little experience. For the first time, I felt vulnerable. The simplest of tasks, tasks that up to now I would have taken for granted, would become extremely difficult, if not impossible. Was this experiment doomed to failure from the start? I decided not to think about it: there was no backing out and anyway, millions of people had heard me talk about it, which added considerably to the pressure I was feeling.

And so as I lay there, covered in oil, full of apprehension, exhausted, stressed and staring at the ceiling, so many thoughts flew through my mind. How on earth had I managed to get to this point in my life and why the hell did I end up embarking on this seemingly impossible mission so publicly?

1

WHY
MONEYLESS?

Money is a bit like love. We spend our entire lives chasing it, yet few of us understand what it actually is. It started out, in many respects, as a fantastic idea.

Once upon a time, people used barter, instead of money, to look after many of their transactions. On market day, people walked around with whatever they had produced; the bakers took their bread, the potters brought their earthenware, the brewers dragged their barrels of ale and the carpenters carried wooden spoons and chairs. They negotiated with the people they hoped would have something of value to them. This was a really great way for people to get together but it wasn't as efficient as it could have been.

If Mr Baker wanted some ale, he went to see Mrs Brewer. After a chat about the kids, Mr Baker would offer some bread in return for some of Mrs Brewer's delicious ale. A lot of the time, this would be perfectly acceptable and both parties would come to a happy agreement. But – and here is where the problem began – sometimes Mrs Brewer didn't want bread or didn't think her neighbour was offering enough in exchange for her beer. Yet Mr Baker had nothing else to offer her. This problem has become known as 'the double coincidence of wants': each person in a transaction has to have something the other person wants. Perhaps Mrs Brewer had discovered her husband was gluten-intolerant and so Mr Baker had been contributing to her lesser half's irritable bowel syndrome. Or that rather than bread, she really wanted a new spoon from Mrs Carpenter and some fresh produce from Mrs Farmer. This was all very confusing for poor Mrs Brewer.

One day, a man in an exquisite top hat and tailor-made pin-striped suit entered the small town. The people had never seen him before. This new chap – he introduced himself as Mr Banks – went to the market and laughed as he watched the hustle and bustle as everyone chaotically mingled and tried to get what they needed for the week. Seeing Mrs Farmer unsuccessfully trying to swap her vegetables for some apples, Mr Banks pulled her aside and told her to get all the townspeople together that evening in the Town Hall, as he knew a way in which he could make their lives so much easier.

That evening, the entire community came, jostling with excitement and intrigued to know what this charismatic stranger in the top hat and beautiful suit was going to say. Mr Banks showed them ten thousand cowry shells, each stamped with his own signature, and gave one hundred shells to each of the one hundred townspeople. He told them that, instead of carrying around awkward beer barrels, loaves, pots and stools, the people could use these shells to trade for their goods. All everyone would have to do was decide how many shells their wares and produce were worth and use the little tokens to do the exchanging. 'This makes a lot of sense', said the people, 'our problems have been solved!'

Mr Banks said he would return in a year and that when he did, he wanted the people to bring him one hundred and ten shells each. The ten extra shells, he said, would be a token of their appreciation for how much time he had saved them and how much easier he had made their lives. 'That sounds fair enough but where will the ten extra shells come from?' said the very smart Mrs Cook, as he climbed off the stage. She knew that the villagers couldn't possibly *all* give back ten extra shells. 'Don't worry, you'll figure it out eventually', said Mr Banks as he walked off to the next town.

And that, by way of simple allegory, was how money came into being. What it has evolved into is far removed from such humble beginnings. The financial system has become so complicated that it almost defies explanation. Money isn't just the notes and coins we carry in our pockets; the numbers in our bank accounts are only the start. There are futures and derivatives, government, corporate and municipal bonds, central bank reserves and the mortgage-backed securities that so famously caused the world-wide collapse of financial institutions in the 2008 credit crunch. There are so many instruments, indices and markets that even the world's experts can't fully understand how they interact.

Money no longer works for us. We work for it. Money has taken over the world. As a society, we worship and venerate a commodity that has no intrinsic value, to the expense of all else. What's more, our entire notion of money is built on a system which promotes inequality, environmental destruction and disrespect for humanity.

DEGREES OF SEPARATION

By 2007, I had been involved in business in some way for nearly ten years. I had studied business and economics in Ireland for four years, followed by six years managing organic food companies in the UK. I had got into organic food after reading a book about Mahatma Gandhi during the final semester of my degree. The way this man lived his life convinced me that I wanted to attempt to put whatever knowledge and skills I had to some positive social use, instead of going into the corporate world to make as much money as I could as quickly as possible, which was my original plan. One of Gandhi's sayings, which struck a chord with me, was 'be the change you want to see in the world', whether you are a 'minority of one or a majority of millions'. The trouble was, I had absolutely no idea what that change was. Organic food seemed (and in many respects still does) to be an ethical industry, so that looked a good place to start.

After six years deeply involved in the organic food industry, I began to see it as an excellent stepping-stone to more ecologically-sound living, rather than the Holy Grail of sustainability I had once believed it. It had many of the problems rife in the conventional food industry: food flown across the world, convenience goods packed in too many layers of plastic and large corporations buying up small independent businesses. I became disillusioned and began exploring other ways to join the growing movement of people world-wide who were concerned about issues such as climate change and resource depletion and wanted to do something about them.

One evening, chatting with my good friend Dawn, we discussed some of the major issues in the world: sweatshops, environmental destruction, factory farms, resource wars and the like. We wondered which we should dedicate our lives to tackling. Not that either of us felt we could make much difference; we were just two small fish in a hugely polluted ocean. That evening, I realised that these symptoms of global malaise were not as unrelated as I had previously thought and that the common thread of a major cause ran through them: our disconnection from what we consume. If we all had to grow our own food, we wouldn't waste a third of it (as we do now in the UK). If we had to make our own tables and chairs, we wouldn't throw them out the moment we changed the interior décor. If we could see the look on the face of the child who, under the eyes of an armed soldier, cuts the cloth for the garment we contemplate buying on the high street, we'd probably give it a miss. If we could see the conditions in which a pig is slaughtered, it would put most of us off our bacon butty. If we had to clean our own drinking water, we sure as hell wouldn't shit in it.

Humans are not fundamentally destructive; I know of very few people who want to cause suffering. But most of us don't have the faintest idea that our daily shopping habits are so destructive. Trouble is, most of us will never see these horrific processes or know the people who produce our goods, let alone have to produce them ourselves. We see some evidence through news media or on the world-wide web but these have little effect; their impact is seriously reduced by the emotional filters of a fibre-optic cable.

Coming to this conclusion, I wanted to find out what enabled this extreme disconnection from what we consume. The answer was, in the end, quite simple. The moment the tool called 'money' came into existence, everything changed. It seemed like a great idea at its conception, and 99.9% of the world's population still believe it is. The problem is what money has become and what it has enabled us to do. It enables us to be completely disconnected from what we consume and from the people who make the products we use. The degrees of separation between the consumer and the consumed have increased massively since the rise of money and, through the complexity of today's financial systems, are greater than ever. Marketing campaigns are specifically designed to hide this reality from us; and with billions of dollars behind them, they're very successful at it.

MONEY AS DEBT

In our modern financial system, most money is created as debt by private banks. Imagine there is only one bank. Mr Smith, who up to now has kept his money under the bed, decides to deposit his life savings, 100 shells, in this bank. Naturally, the bank wants to make a profit, so decides to lend out a proportion of Mr Smith's shells, let's say 90 of them, keeping ten in their coffers in case Mr Smith wants to make a small withdrawal. Another gentleman, Mr Jones, needs a loan. He goes to the bank and is delighted to be given Mr Smith's 90 shells, which he'll eventually have to pay back with interest. Mr Jones takes the shells and elects to spend them on bread, bought from Mrs Baker. At the close of the day, Mrs Baker takes her newly-acquired 90 shells to the bank. Do you see what's happened? Originally, Mr Smith deposited 100 shells in the bank. Now, in addition to Mr Smith's 100 shells, the bank has Mrs Baker's 90 shells. One hundred shells has become 190. Money has been created. What's more, the bank can now lend out a proportion of Mrs Baker's deposit! The process can start again.

Of course, the physical number of shells hasn't changed. If both Mr Smith and Mrs Brown wanted their shells back at the same time, the bank would be in a fix. However, this rarely happens and if it did, the bank would have shells from other depositors to use. The problems start when the bank lends out 90% of all their depositors' shells. The result is that of all the shells in all the bank accounts of this fictional world, only 10% exist! If all the depositors wanted more than 10% of the total amount of shells at the same time, the bank would collapse (a bank run) and people would realise that the bank was *creating* imaginary money.

This system may seem ridiculous but it is what happens today, every day, in every country of the world. Instead of one bank, there are thousands. Instead of shells, we have the world's myriad currencies. But the principle is the same: most money is created by private banks' lending. Our most precious commodity doesn't represent anything of value and the figures in your bank account are mostly someone else's debt, which itself is funded indirectly by another person's debt and so on. Neither are bank runs fictional. Recent bank crises, from Northern Rock in the UK to Fannie Mae in the US, show the inherent instability that comes from basing our financial system on an imaginary resource. The edifice is built on pretence and, as shown by 2009's bank bail-outs across the world, tax-payers inevitably have to subsidise with billions to keep the pretence alive when the system implodes.

DEBT FORCING COMPETITION, NOT CO-OPERATION

In the current financial system, if deposits stay in banks, the banks make no interest and therefore no money. Therefore, banks have a huge incentive to find borrowers by whatever means possible. Whether by advertising, offering artificially low interest rates or encouraging rampant consumerism, banks share an interest in lending out almost all of their deposits. The credit this creates is, in my opinion, responsible for much of the environmental destruction of the planet, as it allows us to live well beyond our means. Every time a bank issues a human with a credit note, the Earth and its future generations receive a corresponding debit note.

It seems we can't get enough of it. According to a Credit Action report published in 2010, there are now 70 million credit cards in the UK; the UK has more 'flexible friends' than people. The average household debt (excluding mortgages) is over £18,000 and to compound the situation, at the time of writing the UK's national debt is growing by an astonishing £4,385 every second. Payback time, in both economic and ecological terms, will inevitably come. Whilst all this money creation is great for the economy, it is not so good for the people that the economy was originally intended to serve. Every day, the UK charity Citizens' Advice helps more than 9,300 people who need professional support to deal with their debts, one person is declared bankrupt or insolvent every four minutes and a house is repossessed every eleven and a half minutes.

In the end, the process of money creation inevitably means the rich get richer and the poor get poorer. Banks lend out money that, by any objective measure, they didn't have in the first place and at every stage, accrue interest and keep the right to repossess real assets if loans are not repaid. Is there any wonder that huge inequality exists in the world?

Let's return to our little town. In the past, at times such as harvest, it was common practice for the people to often help each other out on an informal, non-exchange basis and the people there co-operated a lot more than they do today. This co-operation provided them with their primary sense of security; indeed, a culture of collaboration still exists in parts of the world where money is deemed less important. However, the pursuit of money and humans' insatiable desire for it has encouraged us to compete against each other in a bid to get ever more. In our little town, competition replaced the co-operation that once prevailed. Nobody helped their neighbours bring in the harvest for free any more. This new competitive spirit was partly responsible for many of the town's problems, from feelings of isolation to a rise in suicide, mental illness and anti-social behaviour. It has also contributed to environmental problems, such as the depletion of resources and the climate chaos that currently go hand-in-hand with relentless economic growth.

MONEY REPLACING COMMUNITY AS SECURITY

For most of us, money represents security. As long as we have money in the bank, we'll be safe. This is a precarious position to adopt, as countries such as Argentina and Indonesia, which have recently suffered hyper-inflation, will attest. The boom period the world experienced at the start of the twenty-first century – a bubble inflated by highly-pressurised bank executives – has been punctured. Many politicians, economists and analysts are still not sure if there was only one thorn.

Whilst I've no doubt that we'll make it through this downturn and maybe even a few more, future economic crises will not be so easy to manipulate and stimulating recovery will be harder, as these challenges will be affected by real-world problems. The banking industry is inherently unstable and two of the pillars of our economy, the insurance and oil industries, will eventually take a huge hit from two massive and evolving problems: climate change and 'peak oil'.

CLIMATE CHANGE

Whatever your beliefs about why the climate is changing, it's undeniable that it is. It's also certain that the damage it will cause is going to cost someone an incredible amount of money. In 2006, Rolf Tolle, a senior executive of Lloyd's of London, warned that insurance companies could become extinct unless they seriously addressed the threats climate change poses to their business. Ultimately, there are two scenarios: either the insurance companies continue to cover 'acts of God' (or, more accurately, 'acts of humanity') and drastically increase our premiums to protect themselves – yet still risk extinction; or they stop covering them and the people whose homes and possessions are wiped out pick up the tab, ruining local economies and creating one humanitarian crisis after another.

PEAK OIL

'Peak oil' – a huge subject – boils down to one simple fact: our entire civilisation is based on oil. If you don't believe me, take a look around wherever you are now and try to find one thing that either isn't made from oil (remember plastics are oil-based) or wasn't transported using it. Oil is a finite resource: when it will run out is up for discussion but the fact that it will run out is not. What's more, even before the wells run dry, speculation will push up prices, so that oil will increasingly become unaffordable for more and more people. According to Rob Hopkins, founder of the Transition Network, we are using four barrels of oil for every one we discover, meaning that we are already moving rapidly towards this scenario. To highlight how critical oil is in our lives, Hopkins adds that the oil we use today is the equivalent of having 22 billion slaves hard at work – or each person on the planet having just over three. Oil is the sole reason that we in the West can live the lives we do; lives which are unsustainable in every sense of the word.

Governments may be able to bail out banks during times such as the 2008 credit crunch; unfortunately, we are also approaching what George Monbiot calls the 'Nature Crunch'. As he correctly points out, nature doesn't do bail-outs. Pavan Sukhdev, a Deutsche Bank economist who led a study of ecosystems, reported that we are 'losing natural capital worth somewhere between $2 trillion and $5 trillion every year as a result of deforestation alone'. The credit-crunch losses incurred by the financial sector amount to between $1 trillion and $1.5 trillion; these pale in comparison to the total amount we lose in natural capital every year. As we lurch towards environmental disaster and the economy contracts, will money continue to be seen as security? Or will living in a closely-knit community that has re-learned its ability to work together and share for the common good take its place?

This became apparent to me when I went back to Ireland, to visit my parents, in 2008. In the six years I'd been away from my homeland, working in the UK, the country had changed beyond recognition. The growth that Irish people experienced during the 'Celtic Tiger' economic period had radically affected their culture. Twenty years earlier, when I was growing up at the end of the eighties, it had seemed very different. My memories were symbolised by the street where my parents still live. When I lived there, everyone knew each other; it could take fifteen minutes to get to the bottom of the road on your way to town. Then, out of the eighty houses, only one had a phone. When you wanted to make a phone call, you went to that house (which, like every other house, always had an open door), stuck a couple of small coins on the table and made what was usually a pretty important call. I can remember no more than five cars on the street; if you saw a Mercedes, you knew someone had relatives visiting from abroad.

Now, most people are only interested in getting on their individual property and career ladders. It doesn't really matter what wall the ladder is propped up against, just so long as they are climbing. The street I remember is no longer there; its once-open doors are all shut.

PLANET EARTH PLC

Money allows us to store our wealth very easily and for a long time. If this easy storage were taken away, would we still have an incentive to exploit the planet and all the species that inhabit it? With no way of easily 'storing' the long-term profit that results from taking more than we need, we would be much more likely only to consume resources as we needed them. A person would no longer be able to turn trees in a rainforest into numbers in a bank account, so would have no real reason to cut down a hectare of rainforest every single second. It would make more sense to keep the trees in the earth until we needed them.

Consider the planet as a retail business, whose store managers are our world leaders. These managers of Earth plc are on short, four-year contracts, so they elect to make as much profit as quickly as they can, to give them a better chance of their contracts being renewed. They decide to sell some of the cash registers and shelving, to add a bit extra to the year's bottom line and make the profit and loss account look healthier. It works: the shareholders – us – don't bother to look at the balance sheet and the managers get their contracts extended. The following year, their ability to make money is diminished due to their reduction of important fixtures and fittings and so they have to do the same again, until they have used up every asset they have. In the meantime, the shareholders have voted to re-invest very little of the profit, choosing instead to buy goods with a very short life and of little practical use.

For our planet, it is exactly the same. At the moment, we are liquidising our assets and spending the profits on products with built-in obsolescence. This is a long-term business strategy no responsible businessperson would recommend. In 2009, Kalle Lasn, founder of the influential magazine *Adbusters*, said:

> … we got rich by violating one of the central tenets of economics: thou shall not sell off your capital and call it income. And yet over the past 40 years we have clear-cut the forests, fished rivers and oceans to the brink of extinction and siphoned oil from the Earth as if it possessed an infinite supply. We've sold off our planet's natural capital and called it income. And now the Earth, like the economy, is stripped.

THE DIFFERENCE BETWEEN SELLING AND GIVING

I don't see myself as a hugely spiritual person in the traditional sense. I try to practise what I call 'applied spirituality', in which I apply my beliefs in the physical world, rather than them being something abstract I talk about but rarely practise. The less discrepancy there is between the head, the heart and the hands, the closer you are, I believe, to living honestly. To me, the spiritual and physical are two sides of the same coin.

I do see a non-physical benefit to living without money. When we work for people, beyond what we do for family and friends, it's almost always an exchange: we do something because we get something in return. I believe that prostitution is to sex what buying and selling is to giving and receiving: the spirit in which the act is done is significantly different. When you give freely, for no other reason than the fact that you can make someone's life more enjoyable, it builds bonds, friendships and, eventually, resilient communities. When something is done merely to get something in return, that bond isn't created.

Another major motivation is much simpler and more emotional – I'm tired. I'm tired of witnessing the environmental destruction that takes place every day and playing a part, however small, in it. I'm tired of giving my money to a bank, which, however ethical it claims to be, nonetheless pursues infinite economic growth on a finite planet. I'm tired of seeing families and lands destroyed in the Middle East so that we in the West can fuel our lives on cheap energy. And I want to do something about it. I want community not conflict; I want friendship not fighting. I want to see people make peace with the planet and with ourselves and all the other species that inhabit it.

HOW TO BECOME MONEYLESS

It's one thing to intellectualise the reasons why we should give up money but it's quite a challenge to try and do it. In 2007, I decided to give it a shot. I sold my beloved houseboat, moored in Bristol Harbour, and used the cash to set up a project called the 'Freeconomy Community'. Some might, understandably, call me a hypocrite for using money in an attempt to accelerate its demise. However, I see money in the same way as I see oil: we should be using what we have to build sustainable infrastructures for the future.

I had experience of local trading schemes, such as LETS and Timebanks, in which people exchanged skills and time rather than money. Although I thought these schemes were a really positive alternative to the global monetary system, they still focused on exchange, rather than unconditional giving. My theory was that if you were part of a big enough community, with a diverse enough range of skills, you could help somebody without worrying about what that person could do for you in return. Security would lie in the fact that the community would be there to help any member whenever they needed it. The person whom you help may never help you, and a different person may help you though you have never helped them. The difference between this and the normal monetary system is that one uses figures on a computer screen to calculate our level of security, while the other sees security as the bonds we inevitably build with people when we do something just for the love of it. One system builds stronger communities, the other builds higher fences.

I used the profits from the sale of the boat to pay a web developer to work with me on building an online infrastructure through which people could help each other, not for profit but simply for the love of it. The over-riding aim was for the website

14

to act as a facilitator in enabling people to help each other for free but how best to do this was up for some debate. In the end, I decided that sharing was at its heart; not only did sharing mean that fewer of the world's resources would be used but it would also be a very devious way of bringing people together. Have you ever liked anyone *less* for sharing something with you? Exactly. Sharing builds bonds, reduces fear and makes people feel better about the world they live in. Peace will only come when all the little interactions that occur around the world every day become more harmonious. The whole is made up of the detail.

The Freeconomy Community became a skill, tool and space-sharing website, designed to bring people together and allow them to teach each other new skills, pool resources and eventually be enabled to live a life in which money wasn't the primary factor in everything they did. I called the site 'justfortheloveofit.org', which I felt summed up the spirit of the project. The early success of the website astounded me. The concept behind it was as old as the hills but I suppose its presence on the world-wide web gave it another dimension. Within a year, journalists were using the term 'Freeconomy' to describe the entire moneyless movement.

'BE THE CHANGE'

By early 2008, I felt I was getting closer to understanding what change I actually wanted to be. Having set up a project that successfully enabled people to start making the transition to sharing rather than selling their skills, I decided that if I wanted the world to place less emphasis on money, a decent way to start would be for me to try to live without it, to see whether it were even possible.

In June 2008, I decided that I was going to give up money for at least a year and resolved to start at the end of November, on

International 'Buy Nothing' Day. When I told my friends, they thought I'd gone mad. Why, they asked, was I doing something so extreme (a word that often gets used about my way of living)? But what is 'extreme'? To me, buying a plasma screen television for a couple of grand seems extreme. And given that some of the problems we will face in the future, such as climate change and 'peak oil', are, according to many leading scientists, likely to be extreme, how can we possibly expect the solutions to be moderate?

THE RULES OF
ENGAGEMENT

I come from a very sporting family and before I decided to give up money, I longed to be a mega-rich professional footballer. Football didn't teach me a lot that is relevant for moneyless living but it did show me that every game needs clear rules. Before I did anything else, I resolved to create a set of rules that I could easily explain and which I would follow for the duration of my moneyless year.

Those who inquire about the rules of this experiment mostly fall into two groups. On the one hand are those who are surprised that I would even be thinking of rules. Their logic says that it's my game and I have no opponents, so why not just do as I feel fit at any given moment? On the other hand are those who want to know the answer to every conceivable scenario I could possibly get myself into.

The logic of the first group is somewhat flawed; I do have a couple of opponents. The scarier of the two is my inner demon, which inevitably is weak when exposed to temptations like a lift into town on a wet winter's evening, a dram of whiskey with my friends in front of the fire and, of course, chocolate. Without rules, I know I would sometimes have given in and once I had succumbed, the floodgates would be opened. Knowing your weaknesses will always be one of your greatest strengths. I was less worried by my other opponent: potential critics. When you do something like I was doing, so publicly, you leave yourself open to all sorts of criticism if it fails. I've built up a really good relationship with almost all the journalists I have worked with; the ones I've been fortunate enough to meet have shown a lot of integrity and we have had mutually very beneficial partnerships. However, journalists are paid to get great stories and some publications prefer it juicy. I would be pretty naïve if I thought every producer and editor on the planet was out to promote the message of Freeconomy. One newspaper, whose name I will omit, held a meeting in which the editor asked his team whether or not they should send out an undercover journalist to see if they could catch me spending or accepting money. On this occasion, I happened to know the housemate of one of the editors in the meeting. How prepared they were to create a story one way or another, I'm not sure.

So what were these rules that were going to dictate the parameters of my existence for twelve months?

1. THE FIRST, FUNDAMENTAL LAW OF 'NO-MONEY'

For one whole year I would not be able to receive or spend money. No cheques, no credit cards and no exceptions. Everything I needed or wanted in those twelve months would have to be achieved without either cash or its proxies. I closed my bank account, even though I knew it would be difficult to open a new one with no financial activity for a whole moneyless year.

2. THE LAW OF 'NORMALITY'

Normality is not a word often used in the context of my experiment. The law of normality, however, was my most crucial rule as it provided the intellectual framework for making decisions in the myriad scenarios I'd find myself in during the year. Without this law, I wouldn't be able to undertake my experiment and live a relatively normal life.

If somebody asked if something was within the rules of the experiment, I would ask myself 'what would I normally do?' Take one of the most common questions: 'If a friend wants to cook you dinner one night, would you accept or would you have to barter for it?' This type of question eventually drives you a bit mad. Of course I'm not going to barter with a friend who offers me dinner. In my pre-moneyless days, I would never have offered to pay my friend for the food on my plate; to do so would have gone against every societal norm I was brought up to believe in.

There are some important points to clarify. First, if I offered the same friend dinner a couple of weeks later, or any other friend for that matter, they couldn't refuse on the basis that they were worried I wouldn't have enough food to survive for the week. That would not be a normal excuse. On top of that, if I ever felt people were inviting me to dinner more often than normal, out of concern for my wellbeing rather than just wanting to spend time together, then I would say 'no'. Second, I intended to live off-grid for the year. Living 'off-grid' meant producing all my own energy for lighting, heating, cooking and communicating, and dealing with all my waste. However, that didn't mean that if I were with a friend and they turned on a light or some music, I had to leave the room; that would be ridiculous. I had acquired a laptop computer and mobile phone (for incoming calls only) from people who no longer wanted them; if I visited people far away from home and I needed to charge them, I would use their electricity, if there were no other option, as that is what I would have done in the past. Similarly, if someone came and stayed with me, then I'd offer them any solar energy I had produced. Third, I started the year with a normal amount of food and clothing. To throw everything out on Buy Nothing Eve and start again would go against everything my year was about.

Receiving is very important, as it allows the giver the experience of being generous and kind. Without a receiver there can be no giver and being able to give is one of the greatest gifts bestowed on us. However, it was also vital to keep the integrity of my experiment intact in a very real and everyday way.

3. THE LAW OF PAY-IT-FORWARD

I'd constructed this concept in my head long before a Hollywood film of that title helped me articulate it. The film is about a child whose teacher asks the class to come up with an idea that could change the world for the better. The boy suggests that if one person helped three people with something important, in the faith that each of them would go on and help another three, and so *ad infinitum*, then not only would a lot of caring, kindness and love spread exponentially around the globe, it would also eventually come back to benefit the original giver. Probably at precisely the moment that they needed it most.

I have tried to use pay-it-forward economics to replace traditional bartering. It's about giving and receiving freely. With traditional bartering, both parties agree a 'price' before any work is done; they then carry out what they have agreed until a full exchange has taken place. To me, this isn't much different to money, though it has the benefit of being local and, more often than not, involves things that are important and mutually beneficial. It also creates a more real relationship. However, it lacks an essential spiritual quality: unconditional giving. There is something about unconditional giving that transforms relationships and builds bonds in a way that traditional bartering never could.

When somebody does something for you just for the love of it, with no expectation of anything in return, it is very powerful, especially in the twenty-first century, when we are taught to look after ourselves before everything else. Pay-it-forward is all about unconditional giving. Nature works on this principle: the apple tree gives its fruit unconditionally, without asking for cash or a credit card. It just gives, in the faith that its benefactor will spread its seed further afield, giving the world even more apples.

How does pay-it-forward apply to my experiment? I never agree terms beforehand with people I help; I just help. It's a relationship based on trust. I do it in the faith that my requirements will be met whenever I need (not want) something, whether that help comes from the person I helped or from someone I've never met, whether it comes five minutes after I've helped someone or two years later. Cliché enthusiasts call this 'what goes around comes around'. I believe it's nothing more complex than this: if you spend your time putting more love into the world, then it is reasonable to believe you are going to benefit from a world with more love in it.

Pay-it-forward is a beautiful theory and I believe that if we practised it more fully, the world would be a much friendlier place. We are often too short-sighted and too self-absorbed. We take and we hoard but this creates what is, really, a very false sense of security and abundance. By giving and sharing we could all be so much better off materially, emotionally and spiritually. Not only would we have access to a larger pool of material resources, we'd have a wider network of friends and the warmth that comes with doing something just because we can.

4. THE LAW OF RESPECT

Respecting other people's wishes is an essential part of life and sometimes involves compromise. Even though I planned to live off-grid, I inevitably found myself in other people's houses and workplaces. If I had to do what bears do in the woods, I could have pulled out a spade, dug a hole in the back garden and had a crap; in other words, absolutely shocked my hosts.

However, the point of my experiment was not to annoy and completely alienate the 99% of the population who still use money and sewers; staying steadfastly true to my beliefs in such instances would have been counter-productive. This is what I call my 'law of respect'. I stood up for what I believed in but my focus is on effecting the most positive change in the longer term and on bringing people with me on the journey, if they want to come. If you respect someone else's way of life they are far more likely to respect yours.

5. THE LAW OF NO NEW FOSSIL FUELS

We're soon going to have to make the transition to a world without oil; it is a finite resource and we are using it remarkably quickly. Not only are petroleum-based products incredibly polluting but using them puts pressure on governments to find new sources of oil; a pressure that has, increasingly, resulted in wars around the world.

I didn't want to be responsible for that, so for the whole year no new fossil fuels were to be used in my name. If someone wanted to help by offering me a lift because they thought I was exhausted, I would politely refuse. I would allow myself to hitchhike, as the driver would be going that way anyway. I would only accept lifts when the journey was impossible by foot or bicycle and very sparingly, as the year was not about freeloading on others. I would never hitch the eighteen miles to Bristol to see friends or to get food or wood but I would do so to travel between countries.

6. THE LAW OF 'NO PRE-PAYMENT OF BILLS'

Not only did I not pre-pay any of the normal bills we tend to accumulate, I didn't even have any bills, as I set myself up to be completely off-grid. Setting up my infrastructure didn't happen overnight. In some respects, I'd been preparing for my whole life. However, more practically, I took six months to build up to the year, which I decided would start on 29th November 2008, better known as 'Buy Nothing Day' or, as the last Saturday in November, the day the Christmas buying frenzy officially kicks off in the United States.

3

PREPARING THE FOUNDATIONS

When I first started to think about living for a year without money, I didn't think it would be that difficult. While I always definitely enjoyed having a bit of spare cash at my disposal, it had been a while since my days of rampant consumerism. However, the more I started to explore the rabbit hole that is freeconomic living, the more of a warren it became. Not because it's so difficult in itself but because we in modern western societies have become very conditioned to our comforts and, more critically, have lost many traditional skills. Humans lived without money for a long time – over 90% of *Homo sapiens*'s time on the planet. The problem is that it has become something of a lost art.

One of my first realisations was that there is a huge difference between living on a very tight budget and not being able to spend a single penny. The UK government classifies a household that lives on an income 60% below the median annual income as impoverished. In official terms, anything below

£5,800 is considered poverty and below £5,000 as the severest poverty. Or some kind of living hell on earth. According to government figures, thirteen million people in the UK live in poverty.

Poverty is a funny phenomenon. It is always defined financially and always relative to what other people earn. It is possible to be extremely happy despite having little money and being officially categorised as poverty-stricken. You can also be really unhappy despite earning a high salary. Those who always want something more will always live in poverty, regardless of how much they earn, while those who are content with what they have will always feel they have an abundance. Most poverty in the UK isn't material poverty, it's spiritual poverty, a state of mind in which fulfilment comes only from the pursuit of material gain. Much of the material poverty in places such as Africa stems from the spiritual poverty of the West, as institutions such as the World Trade Organization and International Monetary Fund (IMF) continue to cripple 'developing' nations with debts and restrictions designed to enable western governments to supply the extravagant products and cheap food we, as consumers, demand.

With a bit of organisation, I found I could quite easily live on £5,000 a year, even after rent. The problems begin when you cannot use money at all, turning what would normally be a small purchase into a huge undertaking. Let's say you live on a tiny wage of £50 a week and your pen runs out. Pens are, monetarily, cheap; almost anyone can run to the nearest shop and pick up a new one for 25p. Without money, this becomes an entirely different prospect. It doesn't matter if pens are unbelievably cheap, it doesn't matter if they drop to 5p; without money, you simply cannot buy one. Instead of spending the equivalent of two minutes' work at the UK's minimum wage, you'd have to spend three-quarters of a day making a new pen from inkcap mushrooms. This is the difference between living frugally and living completely without money. This reality scared the hell out of me.

DECONSTRUCTING MY CONSUMPTION HABITS

Journalists and reporters, it seemed, sensed the enormity of my experiment long before I did. In the early stages, often the first question they asked was 'how are you going to do this?' They hoped for a short sound-bite they could fit into their interview or article. Yet how do you explain succinctly how you are going to live without money for an entire year?

I found the best answer to this question was to be honest about how I had prepared. When I decided to live for a year without money, the second thing I did, after formulating my rules, was to get a notepad and list every single thing I consumed; as it stood, there and then. I called this my 'breaking-it-down' list. To structure my thoughts, I categorised my list into food, energy, heating, transport, entertainment, lighting, communications, reading, art and so on. The list eventually took up half the notepad – and that was the list of someone who considered himself quite a moderate consumer. I'd shudder to think of the list an A-list celebrity would come up with. I worked my way down the list, trying to figure out how I could acquire all the things I would normally need in ways that didn't involve money. It became clear, after just a couple of pages, that most of the stuff would involve me having no more than one degree of separation from what I consumed; either I would make it myself or know the person who produced it.

This was a perfect starting point. It provided me with lots of really useful information with which I could make decisions. How many new skills would I have to learn either before or during the experiment? How much was the necessary infrastructure going to cost? How much time would each activity take? As this year was about consuming less and having a closer relationship with the things that remained, my list-making enabled me to establish my basic level of subsistence, the things I really couldn't do without, and my priorities for the rest.

One of the greatest things about this process was that it forced me to ask myself how important each item was. I love bread; it's a deep-rooted addiction. The 'breaking-it-down' process made me realise that to have bread I would have to get my grain, take it home in my bike's trailer (which always takes a bit longer on busy roads) and grind it into flour using a hand-cranked grain mill. I would have to make my sourdough starter and (for the first batch) wait five days. During these five days I would have to make a cob oven outside. Once that was made and fit for use, I would have to fire it up, then look after it constantly for a couple of hours as my bread cooked. By which time I would probably be too tired to eat the delicious loaf I had spent a week preparing.

I subscribe to the 'Permaculture' ideology. Permaculture is about creating human habitats and food production systems by designing models which mimic natural patterns. These models not only eliminate almost all waste and save lots of energy but also save a lot of work. Whilst I certainly wouldn't call myself lazy, I don't believe in using more kilojoules of energy to make food than that piece of food will supply, or it would make more sense to lie down and read a book. However, there is always a middle way. The listing process made me realise that, if I wanted bread, I was going to have to come up with a new solution. And I did. I decided that although I loved bread, it would have to be a treat. Instead, I would sprout the grains. This means sprinkling a layer of rye grains along a couple of stacked, perforated trays and rinsing them with water twice a day until they sprout. This only takes five minutes and so is much less effort, for more nutritional gain, than making bread. Although not quite so pleasing to taste and smell!

This is just one example from a list of hundreds. Another benefit of the list was that it enabled me to figure out how much I would have to save and then spend to create the infrastructure necessary to make this year happen. It may sound ironic, or even contradictory, to hear me say I had to save and spend money to make my year without money happen. But I never said I wanted humanity to stop using money tomorrow, any more than I would like to see humankind stop using oil next week. Much as I would love to see both happen one day, currently it would cause catastrophe, as our entire infrastructure is based on the abundance of both. I view money in the same way as oil; if we insist on using it, let's at least stop using it for non-essential or destructive goods and services. Let's start using both these resources to build a new infrastructure that will enable us to be truly sustainable in the long term. For me, it is not about revolution, but about evolution, transition and transformation.

To get the basic infrastructure I felt I required to live without money, it looked as though I was going to have to save around £1,600 in four months, at the same time as putting in the work needed for it all to come together in time. This figure was based on me buying everything (such as a home) either new or at the higher end of what it could cost if I were to buy it second-hand, as I didn't want my year to fail before it had even started. However, I had no intention of buying everything. My plan was to see how much I could get for free by using other people's waste, which was completely in keeping with the spirit of the project. This would be time-consuming but possible.

The research was complete and the lists were made. It was time to start getting all the listed items together before the end of November.

SETTING UP MY INFRASTRUCTURE

SHELTER

When you are trying to figure out what you need to survive, the first things you think of are the basic necessities. Top of that list was shelter.

When I first decided to live for a year without money, I had no idea how I could avoid paying rent. I knew I couldn't live in a normal house, as that would involve spending substantial amounts of money. I would have to acquire some sort of shelter and put it somewhere I could use for free. In the beginning, I was prepared to consider anything – a tent, a yurt, a caravan, a tipi; I didn't care. Obviously a tent would not be a great place to spend the winter but I was so determined to start that I seriously considered it. In my 'breaking-it-down list', I had budgeted £500 for housing, which is 0.2% of the average house price in my area. This could get me an amazing tent, a ridiculously small caravan or the left-hand side of a yurt. And I wasn't even sure if I could afford that. So I decided one morning to try my luck and post an advertisement on Freecycle: 'WANTED: any type of living structure – tent, yurt or caravan'.

I'd included the caravan almost as a joke. You can imagine my surprise when I got a reply from a woman who said she had a caravan that she would be more than happy to give me. My first reaction was 'too good to be true' and that it was probably junk and falling apart. It turned out to be nothing of the sort, just a perfectly decent caravan, which, because it was over ten years old, was no longer allowed on caravan parks. The owner couldn't sell it without doing more work than she was prepared for and storing it was costing her £25 a month. I asked her what she wanted from me to take it off her hands. She handed me the keys with a smile on her face and said it was mine. This was a major success – not only had I a much bigger, warmer and sturdier home than I was expecting, it also meant I could wipe £500 off the money I was going to need to set up my year. Likewise, I hadn't had to buy something new, which was equally important to me.

GETTING THINGS FOR FREE

There are two fantastic online tools that match up people who have stuff they no longer need with people who could use it: **Freecycle** (www.freecycle.org), which has groups all over the world and **Freegle** (www.ilovefreegle.org). Not only do these projects keep perfectly usable products out of rubbish dumps (Freecycle keeps four million tonnes of useful stuff out of the ground each year), they also reduce carbon dioxide production, as the recipients don't have to pull new products through the supply chain; products that inevitably contain a lot of embodied energy (the amount of energy it took to make and distribute the product).

Through these online systems, not only can people advertise things they no longer want but others can also stick up 'wanted' notices. If anyone in these groups has it and is happy to send it to a new home, they can contact you at their leisure.

Another, more traditional, way of meeting your material needs is either to go to your local recycling depot or scour your neighbourhood for stuff that people leave outside their front gate or in a skip. It is amazing what you can find – many of the things I use were destined for landfill.

Now that I had my home, I had to find somewhere I could park it for free. Renting in any city isn't cheap; a few weeks earlier I had worked out that the first seven working days of every month went straight to my landlady. I realised that now I wouldn't have to pay rent, I would be free to volunteer for projects I really believed in and wanted to support with my time. I had an idea – why not work for some volunteer project that had a piece of land where I could pitch my caravan, without it interfering in their operations?

I made a list of possible organisations. One stood out from the rest: Radford Mill Farm, near Bristol. Although it was further away from the city than the others, it was a project I really wanted to help. They needed me to work more days than the others – three days per week; the equivalent of having to work the first twelve days of every month for the roof over my head. As a reporter pointed out, this compared badly to the seven days, on minimum wage, I would have to work to pay the rent for a room in Bristol. However, this was the kind of thinking I was trying to move away from. As part of the deal with the community of people who ran the farm, I agreed to work three nine-hour days, though this was a flexible and informal arrangement. When you are doing something you want to do, you don't mind doing a bit extra when needed. I would help grow food and manage the land (the hedgerows in particular), along with everything else it takes to run an organic farm, such as cleaning and composting. Because the farm had 100 acres, there was plenty of room for me to grow my own food.

The only difficulty I had with my new home was that it didn't have a toilet. To solve this problem, I decided to build a compost toilet. This isn't much different to a normal toilet, except it doesn't have a flush and so doesn't waste lots of water. It also provides useful fertiliser. There are two approaches to building a compost toilet: one is to spend three days making a beautiful structure, on stilts, that will give you plenty of comfort and the ability to easily produce 'humanure' (yes, it is what you think it is!). Humanure is very good fertiliser if made properly and when it is ready to spread on the land, the stilts make access to the composted waste easy. The other option is to spend half an hour making a 'modesty structure' from three pallets and another ten minutes digging a three-feet-deep, foot-wide hole in the ground. This way, you get yourself both a mobile toilet and shower unit. Here's how it works – you crap in the hole, then clean yourself using your preferred method. Once you've finished you scoop some of the soil you dug out over the top of what you've produced, so that it doesn't create unpleasant aromas or attract mice and rats. When the hole is full, you dig another one and move the modesty structure. The structure can also cover you as you shower. Not that I cared, but there was a public right of way next to my caravan and I think the last thing some poor dog-walker needs is to see a naked Irishman shivering under a tree on a frosty winter's morning.

While my compost loo is a bit of a joke to some folk who come to visit, to me it is the symbol of everything I am trying to do. The compost loo represents sanity and respect, not just for the environment but also for our fellow human beings. I really believe that until we stop polluting the rivers that give us life, nothing will change. Change will only come when we learn a greater respect for nature. For me, the 'normal' toilet represents everything insane and destructive in the world. We take clean water and defecate in it. Human shit is great for the soil but terrible for the water supply. To make it clean again we build large water treatment plants, blast the water with all sorts of chemicals and then put it back into the system. This not only takes lots of energy, it also means that we drink water that once had shit and now has chemicals in it. It is absolutely crazy and illustrates perfectly how our current way of living treats the environment with disdain.

During the years that India was peacefully fighting for independence from Britain, Gandhi used the spinning wheel as the symbol of the movement. He knew that real independence would only come when India had economic independence and that India, as a nation, needed to earn the right to be free. The Indian people's refusal to buy British cloth and their decision to restart producing their own brought them independence in the end.

Now that I had set up a relationship in which I could have a roof over my head without the need for money, my next concern was how I was going to make my own energy.

ENERGY

Living without money and off-grid meant producing all my own energy. There would be no bottled gas, no disposable batteries and no hook-up to the national grid. If I weren't off-grid, I would either have to pay a bill or live completely without energy; neither of which was feasible.

I had wanted to go off-grid for a long time, as I believe we should take responsibility for looking after our own energy and waste-disposal needs. This gives us a better appreciation of what we consume. On top of that, you couldn't imagine a more wasteful energy network than the national grid. While it's great to fill your kettle half-way and change your lightbulbs to more energy-efficient ones, these actions seem insignificant when you consider the grid's larger, systemic, problems. According to Greenpeace, 'our centralised model of production and transmission wastes an astonishing two-thirds of primary energy inputs, requiring us to burn far more fuel and emit far more carbon dioxide than necessary'. Two-thirds of all the electricity that is produced and fed into the grid is lost before it even gets to your sockets! This is the idiotic system that governments are clinging on to and as they don't look like they are going to do much about it, it is up to us as individuals to take the lead.

How easy it is to go off-grid depends on how much you have to spend. If you have £10,000 to make it happen, it's easy. But if your budget is £350, it's very different. The challenge of proving that you don't have to be rich to live in a more environmentally-friendly way was a huge motivation for me. People always tell me organic food is only for those who can afford it but it's not. If you care about what you put in your body, keeping chemicals, pesticides, synthetic fertilisers and additives out of our food is more important than having a couple of hundred television channels. In 2008, I managed to eat a 100% locally-grown organic food diet on a part-time job's minimum wage. In the autumn of that year, this diet became the trigger for a successful national campaign, *Eat the Change*, in which thousands of people across the UK pledged to do the same for one whole week.

There were a few basic things I would definitely need: a cooker, something to heat the caravan that used waste local materials, a light, a wind-up torch, a shower and an energy source for the laptop and mobile phone – my 'transitional tools' – so that I could communicate and document the year. The most important was obviously the cooker; without it, I would be eating raw food for the year. The first idea that sprang to mind was to combine heating and cooking and consequently use half the energy and effort. However, this would have meant that during the hot summer months, cooking my food would have meant cooking myself at the same time. So I came up with another solution.

A few weeks earlier, I had organised a workshop with a great friend of mine, Chris Adams, as part of the Freeskilling evenings that our local Freeconomy Community group puts on each week. On these evenings, one member of the community gives a free demonstration of their particular skill to anywhere between 15 and 150 other members. This workshop just happened to be on how to make a rocket stove: a very fuel-efficient cooker made from recycled materials. Chris was just about to leave on an overland around-the-world trip and he had a lovely rocket stove he wasn't going to need. Knowing it couldn't possibly find a home where it would get more use, he kindly offered it to me. I now had shelter and a cooker, both stuff other people didn't want any more and both for free.

HOW TO MAKE A ROCKET STOVE

THINGS YOU WILL NEED:

Two catering-sized olive cans (or similar-sized cans). Ask at your local delicatessen
An elbowed flue pipe, four-inch minimum diameter
Tin snips
A tough pair of gloves
Insulating material (such as vermiculite or ashes)

INSTRUCTIONS

1. Carefully cut the entire bottom out of one of the cans with your tin snips. Then cut a round hole, the size of your cooking pot, in the top of the same can. Don't leave any sharp edges.
2. Cut the entire top off the other can.
3. Cut a hole, the size of your flue pipe, in the front of the second can.
4. Make a small slice in the bottom of the first (cooking) can so that you can bend its bottom edge around the top of the other can.
5. Now place the flue pipe in the hole and position it up through the bottom can, leaving at least two inches around it free, so that you can fill the sides with your insulating material. This will simultaneously keep it steady and reduce heat loss.

6. Place the other can on top of this and again fill up the sides with insulating material.
7. Put your pan on the top, light some small bits of kindling at the bottom of the elbowed flue pipe and get cooking!

Once you light the wood at the bottom it shoots a flame up the flue pipe to the bottom of the pan, hence the name 'rocket' stove!

Figure 1 'My cooker!'

To heat the caravan, I decided a woodburner was the only option. I could burn waste wood, which is very environmentally friendly. Rotting wood produces methane, a greenhouse gas more than twenty times as effective at heating the Earth's atmosphere as carbon dioxide. Strange as it may sound, it's better to burn wood than let it rot. That is not to say you should clean up a forest floor; urban wastelands are much better foraging grounds, as the wood there does not form part of a natural ecology. I also knew I could coppice woodland on the farm, which is a part of good land management.

After weeks of research and enquiries, I got a tip-off that a guy in a local squat made woodburners from recycled gas bottles, bike parts and scrap metal. Squats often get a bad press and people believe those who live in them are freeloaders who contribute nothing to society. The reality is often opposite: those who live in these previously unoccupied buildings usually give as freely as they receive. And my woodburner proves the point: Gavin, the squatter, whom I've come to know, knocked me up a beautiful one from waste materials for £60. When you consider it took him over a day to make it, this was an absolute bargain.

Ideally, I wouldn't have needed electricity at all; laptops and mobile phones aren't exactly necessary for survival but a major part of my year was about communicating my experiences with anyone who was interested. You can't make electricity from nettles and rosehips, so some element of environmental damage was inevitable. I had a number of options – wind power, solar power or wind-up generation. I just had to choose the option that would minimise the damage. During the British winter, wind power is the best choice, whilst solar power is the best in the summer. Wind-up is very laborious at any time of year, though it has the benefit of not being reliant on the weather. Ideally you would have all of them; diversity is the key to covering all eventualities. Having weighed up all the options, wind power was my first choice but I really couldn't afford anything that suited my needs. So I decided to go with solar power. Trying to find a second-hand solar panel is like trying to find a sober Irishman on the seventeenth of March, so I went against everything I believe in and bought a new one, in a half-price sale, for £200. There is a lot of embodied energy involved in solar power production; the minerals and materials used to make the panels have to be mined, processed and shipped around the world, so I wasn't overly happy.

Because I was quite a distance from the nearest streetlight – or indeed any night-time light other than the moon – a torch was essential. To my delight, I found a wind-up torch – a present from a journalist – lying forgotten in my old backpack.

To wash myself, there was no option that really excited me. The easiest thing was to buy a solar shower. When I say easy, I mean it was easy to buy, because it certainly wasn't easy to stand under it at seven o'clock on a frosty winter morning. I picked one up for £5, again new, which seems cheap until you realise it really isn't anything more than a thick black plastic bag with a bit of hosepipe at the bottom. However, they are quite effective in the summer. You leave them out in the sun during the day and because they are black they absorb the sun's heat. On a warm English summer's day, they can bring the water inside up to well over 20°C. After buying it I got a bit of 'the guilt' – I really should have made one. And if time allowed, I decided, I would make myself a hot tub if I could get my hands on an old bath via Freecycle. My water came from various sources. For washing I used river water, but for drinking I mostly took tap water from the farm, as tests by a local scientist had shown the river to be contaminated with various pollutants. And every now and then, after prolonged rain or snow, a spring rose at the bottom of the valley, which I used whenever I could.

Getting my new home and kitting it out for off-grid living had cost me a total of £265, flying in the face of those who say that environmentalism is only for the wealthy middle classes with nothing better to do. I fully accept that I was very lucky and it did take me quite a bit of work to do it so cheaply. But even if you stick on another thousand pounds, it would still be within the reaches of most in western society, given how much we all spend on furniture alone. Now that the homestead was as good as sorted, my next priority was to figure out how I was going to feed myself.

FOOD

In the West, our general appreciation of anything related to food – growing, foraging and perhaps even cooking – has decreased significantly since the Second World War. The last generation of people who had to grow food to survive are elderly. And although there has been a welcome recent interest in growing our own food, many people today have little idea where their food comes from, beyond the supermarket. A good friend of mine who takes kids on educational walks around organic farms in Bristol once asked a group of ten-year-olds, 'does anyone know what this is?' while pointing at some rosemary in a herb garden. After twenty seconds one hand went up and proclaimed that it was corned beef. He wasn't joking; worse still, nobody laughed. Given this, it should come as no surprise when I tell you that one of the first things people ask me when they hear I am free of money is 'what on earth do you eat?' A lot of people today think that food just comes from the supermarket.

The reality couldn't be further from the truth. To start with, Mother Earth doesn't charge a penny for her fruits. Money is our invention, not hers, though to listen to many people you'd think it had the same status as water, food and oxygen. There is food for free everywhere. You just need to know where to look and what to look for.

There are four legs to the money-free food table. The most exciting is foraging, which originally meant wandering in search of food and provisions, though it is mostly used these days to describe the act of picking and eating wild foods. I am not much of a forager. It's not that I don't enjoy it but it takes a lifetime to learn and I am a relative novice, although my knowledge is much greater than it was. Necessity is a great educator and I am also lucky enough to have a number of foraging friends who have helped me to learn. One, Fergus Drennan, who sprang to fame as the BBC's 'Roadkill Chef', is one of the world's foremost foragers. And two of my old allotment neighbours are Andy and Dave Hamilton, self-sufficiency gurus and co-authors of *The Self-Sufficient-ish Bible*.

Foraging in modern society can never be about getting all your food from the wild but it can be a great supplement. In my ideal world, we would all forage for the majority of our food. However, given that there isn't a lot of wild left and that the population of the UK is now more than 61 million, there isn't enough for everyone. The food you do get from foraging is highly nutritious; it's also vibrant and alive and so much fun to find and pick. What's more, the whole experience is absolutely free and anyone can do it, though I would always recommend not eating anything unless you are sure it is safe. Complete novices should start with simple things like apples, blackberries and nettles and work up their knowledge as they go along.

The second leg of the food-for-free table is what I call 'urban foraging'; using other people's waste. The news media like to portray this as jumping into somebody's bin for a bit of 'skipping' (sometimes called bin-diving or bin-raiding), though the reality is very different. Skipping definitely does have its place, though it is getting increasingly difficult. The problem with it – which is also the great thing about it – is that you never know what you're going to get until you go. You can quite easily come away with a lot of your week's supply but, nutritionally, I wouldn't recommend eating only waste food. You very rarely get good organic, fresh produce and any diet lacking in that is unhealthy in my eyes, given the amount of oil-based pesticides, herbicides and synthetic fertilisers sprayed on conventionally-farmed fruit and vegetables. But skipping is perfect for products you cannot grow or forage for without a lot of processing and the right tools. I prefer to build relationships with those businesses that throw perfectly good grub away because they want to have the reputation of only selling the freshest food. They often have to pay for its disposal and I find that if you approach them in the right manner, they are more than happy to give you their waste. When it comes down to it, very few people want to chuck out good food, especially considering that almost half of the world's population suffers from not having enough.

The remaining two legs of the table are the two ways of acquiring fresh, local, organic produce and grain without money. The obvious one is to grow the food yourself. It's extremely difficult to make a profit from growing food organically, on a small scale, as supermarkets have completely altered what the public perceives to be a normal price. The few farmers that do are certainly not in it for the money, as there are much easier ways to make a living; most do it because they are passionate about growing chemical-free food in a way that respects the long-term health of the soil. However, there is nothing to stop you growing food yourself. It seems insane to me that a small-scale farmer should spend long hours growing food, then sell it at minuscule wholesale prices and use the profits to buy their own food at much more expensive retail prices.

It is really difficult to meet all your own food needs by yourself, unless you are part of a community that grows and eats its food communally. This is where the last leg comes in – bartering. Bartering can either be an exchange of food, especially in the summer when many people have gluts, or an exchange of skills for food or skills you don't have. I like to do it informally; work hard for somebody during the day and at the end of it receive a non-negotiated amount of food.

Some people say this sounds very risky but I've yet to come away feeling ripped off. Sometimes I tell people that I worked all day for a fifty-five-pound bag of oats. They usually think I am crazy; you can buy the same bag for £20 and I've done nine hours' hard labour. But these people are thinking conventionally. I think we need to be more aware of the real cost of food. Those fifty-five pounds of oats should never cost £20. If I had to plant, weed, water, harvest and roll that much oats, it would take me about sixty hours. Therefore, I get sixty hours of work for only nine, which I think is a great deal, as does the person I help. That's the beauty of it. These relationships form much tighter friendships between people and I believe can play a crucial role in our efforts to rebuild communities around trust; relationships in which friendships, not cash, are seen as security.

I spent four months building relationships, either with the land on which I live or with the people of my local community. I learned where the best skips were, which businesses had waste food, where I could find wild foods, who I could help and some of the skills I would need to grow my own food. Strength lies in diversity and the more sources of food you have, the more chance you have of surviving when one lets you down.

Nevertheless, as some of the people with whom I built relationships were eighteen miles away in the city, my next challenge was to set myself up with a means of transport.

TRANSPORT

There are two main forms of free transport, although they often have hidden costs. Walking is completely free if you are prepared to walk barefoot or to make your own shoes. Otherwise, just like the pen, it is extremely cheap but not totally free. I learned how to make flip-flops out of old car tyres, spare fabric and used bicycle inner tubes: I cut the shape of my foot out of the tyre, clad it in some comfortable material, preferably hemp, and used the inner tube as the bit I put my toes around. Walking is my preferred mode of transport. These days even cycling seems too fast. When you are walking you can hear the birds sing, you can check out the plants around you and it's a great way of relaxing and exercising. But walking takes time and, given the time restraints inherent in money-free living, I decided that unless I was really ahead of myself, I would always use my bike.

The second possible form of free transport is the bicycle. Obviously, bikes are made from parts and if one breaks you need to replace the part or fix it. That's not to say you can't do it without money; you just need to build a relationship with someone who has access to bike parts, which may involve bartering. I get my bike parts from a couple of local shops that have to chuck out whole bikes because one major thing is wrong, even though most of the bike is perfectly fine. Because they can't sell a used part, such as a brake pad, they would otherwise have to send the whole thing to landfill.

As part of my year involved me using other people's waste, I had to find a way of carrying items on my bicycle. I had budgeted only £160 for everything transport-related, which didn't leave a lot for a trailer. The cheapest one I could find, which was small and not very sturdy, was £80. I went to a few second-hand bike stores, where I found one of those carriages that parents normally carry their kids in. It was only £70 and was quite a bit bigger than the trailer. Knowing that my chances of getting one from Freecycle were very slim, I bit the bullet and bought it. I also got a good pair of waterproof panniers for £50. For lights, dynamos don't require batteries; a friend who is petrified of cycling nowadays gave me one.

COMMUNICATIONS

While it is great being able to communicate with people, especially when what you are doing may be a resource for others, it's not exactly necessary for survival. Even if I had been cut off from all or most forms of communication, it wouldn't have meant I couldn't live without money; I just wouldn't be able to share the experience as effectively.

Two things that have completely changed the way we live since the 1990s are mobile phone technology and the Internet. I have a love/hate relationship with the phone; when socialising, I prefer to just call and see people. But I knew that if I wanted to communicate about my year without money to the world I was probably going to need a phone, at least for the first few weeks. How to run a mobile without money was an issue. I had a 'pay as you go' phone – with no contract and no bills – but I thought I might get cut off if I didn't put credit on it every three months. Friends thought I should put lots of credit on beforehand but this wouldn't have been living without money and would certainly have violated my 'normality' rule. So, I put no credit on and hoped for the best. This meant I could only receive incoming calls but it was better than nothing.

The farm has a landline phone and they were more than happy to let me receive calls for interviews (radio stations don't like interviewees to use a mobile phone because of the poor sound quality). They were also happy for me to make some calls, as I was working so much, but I didn't feel that spending their money should form part of the experiment. The farm also had some WiFi flying around, which residents at the farm were already using. This meant that I could easily keep up with my commitments to the Freeconomy Community.

EVERYTHING ELSE

My main goal in the run-up to the start of my year was to ensure that I had thought about and prepared methods of providing shelter, food, heating, energy, transport and communication. There were many other areas of everyday living I could have thought about and several things I couldn't possibly predict I might need. However, I decided that I was just going to have to deal with everything else as it arose. There really is only so much preparation you can do and I decided to put faith in the old maxim that 'necessity is the mother of invention'.

With that, I put away my list, relaxed my shoulders and resolved to curtail my use of money as much as possible immediately, as a trial for the year itself. I thought it would be wise to get in a bit of practice before I started what was going to turn out to be a very public experiment.

4

BUY NOTHING EVE

THE WEEK BEFORE

When you are preparing for a momentous change in your life, the reality often doesn't kick in until a few weeks beforehand. Then you start thinking about how it is really going to affect your life, wonder why the hell you decided to put yourself in such a position and occasionally, inevitably, ask yourself whether you can get out of it.

It was only during moments of complete exhaustion that I felt like that. I decided to launch my moneyless year by putting on a 'Food for Free Freeconomy Feast' in Bristol. I aimed to make a free, three-course, full-service meal solely from waste and foraged food, for as many people as I could. The problem was I was already quite stressed about everything else I had to do in the run-up to the start date. And here I was taking on a mammoth mission right at the beginning; making what was already going to be a demanding day even more difficult. I also decided that it would be a good idea to start living the no-money life a week early, giving myself the luxury of a trial run, with the idea that before Buy Nothing Eve, I could acquire any infrastructural requirements I had overlooked.

It turned out that this wasn't even close to being a good idea. I had so much to do in the city that week that living the slow life in the country was impossible. I abandoned my trial run after just two days and hoped that I hadn't overlooked anything too critical. I decided to stay with Claire in the city for the rest of the week, giving me a chance to spend some time with her in normal circumstances, which I felt was particularly important as we had only been seeing each other for a few weeks. Spending a few days in the city was a good idea; it bought me some time, took some of the pressure off and gave me a chance to catch up with myself. I had an instinct that the enormity of the year ahead would soon make more such chances almost impossible. This was when I really started to feel the pressure, with moments when I definitely thought about packing it all in and having a normal life, in which I got to spend time with friends, go for drinks and holidays and maybe even have some time off every now and then.

Preparations for the free feast were going reasonably well. I'd scored 200 pounds of vegetables, about to be composted, from a local organic produce wholesaler. The problem was we had no idea how many people would show up and, given that we were organising a free meal, cooked by chefs like Fergus, there was a chance it was going to be very popular. We needed much more produce.

By the evening of Thursday 27th November, I was mentally exhausted. I wanted the year to start and to get back to living again. I decided to take the next day off, catch up on some reading and tie up a few loose ends. Oh, and go for that last pint.

CASHLESS COMMUNICATION

Before my moneyless year, I faced a difficult decision; whether or not to use two of the products – which most people would classify as luxurious – of financially-fuelled industrialisation: a mobile phone and a laptop computer.

It was a dilemma. If I decided not to use the tools that would enable me to communicate my experiment to the world, I risked being criticised for running away, looking out for myself and not contributing to society in any way. I also knew that if I did use them then I'd also be criticised, as I would be speaking out about money and industrialisation using two pieces of technology that were reliant on both, which could be perceived as being very hypocritical. I decided to use them. If using them meant that I could let even one person know about moneyless living, that alone would be worth the accusations of hypocrisy.

Communicating without cash is obviously never going to be as convenient as with it, but it is still certainly possible. Communicating with those who live nearby has always been free; it just involves getting together. I've found it really beneficial to have been forced back into this situation. However, as cheap travel has enabled us to have family and friends dispersed across the world, we have a huge need for technological communication.

For email, there are quite a few options. You can usually get it for free at your local library, which doubles up as a great way of sharing a computer. If you have your own computer and Internet access, you can use Skype (www.skype.com) to make completely free computer-to-computer 'phone' calls with anyone else in the world who also has Skype. Many websites (such as www.cbfsms.com) allow you to send free text messages but be careful which you choose; some of them cost the person who receives the text message, which is hardly the point.

However, for all these, you need a computer. If you know how to put one together, you can easily get all the parts from Freecycle. Once you have the hardware set up, you can use Linux, a piece of free and open source software, as your operating system and OpenOffice for your spreadsheet, presentation and word processing needs. OpenOffice is compatible with all Microsoft Office applications. Linux also has the added benefit of being really secure, so you don't need to fork out on expensive security and anti-virus software.

Failing that, get two cups, a very long piece of string and ...

BUY NOTHING EVE, 28TH NOVEMBER 2008

Having given myself the day off, I was finally starting to feel more relaxed and looking forward to the day ahead. My schedule was supposed to look something like this:

07:00 Wake up, have some breakfast and read for a bit.

09:00 Meet Fergus, drive in his van to the wholesale food market to see if we could get some waste vegetables for the feast.

11:00 Go into town, print leaflets for the feast and pick up any infrastructural stuff I felt I still needed.

13:00 Lunch.

14:00 Back to bed for an afternoon nap and some reading.

17:30 Dinner.

19:00 Meet my friends Chris and Suzie for a few farewell drinks before they took off on an around–the–world trip over land and sea.

22:00 Bed.
22:01 Read.
22:02 Sleep.

Life regularly indulges its annoying habit of not letting things go to plan. Instead of relaxing before the most bizarre day of my thirty years on this planet, my day turned out something like this:

07:00 Wake up.
07:35 Wake up again and disable the snooze function on my phone. Have some breakfast and read for a bit.
09:00 Meet Fergus. So far, so good.
09:30 Realise the battery in Fergus's van has died overnight. Shit. That is not good. Blood pressure is rising, anxiety is increasing and need to hide in a cabin in the woods is becoming stronger.
09:35 Realise there isn't a chance of being able to get the battery charged and make it to the wholesale market before it closes at 11 am.
09:50 As the market won't re-open until five o'clock tomorrow morning, decide to give up on it for the day and leave getting 60% of our food requirements until the morning of the feast. This is obviously a very risky strategy. I am slightly anxious.
10:00 Go back to Claire's house to check my emails.
10:15 Read an email from a journalist I spoke to earlier in the week, who says her article has gone into the *Irish Times*. I read the article; it's far too kind to me but I have a sinking feeling I know what is going to happen next.
10:20 I get a call from BBC Radio Bristol, asking if I could do a live interview at eight o'clock the next morning. I say 'yes' despite knowing I have also got to find about 300 pounds of waste food, somewhere, at around the same time. I realise my

story is probably now on the BBC news feed, so I put away the book I planned to read.

10:25 BBC Breakfast News phones. Can I come in for their morning show? They say they'll pay for taxis up and down to London. I tell them I think they've missed the point, so they agree to send one of BBC Bristol's satellite trucks to do it live. I tell them I have already agreed to be on Radio Bristol at the same time; they say that they'll look after that. Why do they want to speak to someone who, as yet, has achieved absolutely nothing?

10:30 BBC Radio Bristol rings to say that I can do their interview from the same truck.

10:35 South West news feed phones for some more details. This is both a good and a bad thing. It means my message will get lots of publicity over the next two days; it also means I'm going to get very busy.

10:40 BBC Radio Wales phones and asks for an interview tomorrow morning too. Nice. I now have to do three interviews *and* find 300 pounds of vegetables for free before eight o'clock on the first morning of my year without money. But I never refuse interviews, especially live ones, as each one is a chance to get the message out.

10:45 My story is obviously on all the news feeds. Sky News is now on to it. They just want some quotes for their website, which is often the feed for the news stories on the Yahoo homepage. Good job Fergus's van wasn't working.

10:50 Start replying to as many emails as I can but they are coming in more quickly than I can type.

10:55 *Newstalk*, one of Ireland's major radio stations, calls. I do an interview. All over in ten minutes.

11:05 A journalist from *The Independent* newspaper wants to do a story and calls to find out more. She talks forever and meanwhile Claire's phone is constantly ringing. It was a bad idea to give out the number to the news feed agency. Now I have two phones on the go, non-stop. I tell the journalist I need to go.

11:15 Someone phones about making a short documentary for Korean TV next week. Apparently the Koreans have gone money-mad over the last decade and they think I'd be a good, thought-provoking story for their viewers. They'll probably just make me look mad, but I say yes. Today I have decided, like Danny Wallace (author of *Yes Man*), to say 'yes' to everything and see where it gets me. It would be a good day to ask me for a one-year interest-free loan, if I had any money.

11:20 Sky News Radio phones for a pre-recorded interview. I say something about my year not being nearly as extreme as the entire news media being owned by so few people. I have a feeling they won't use that bit.

11:25 Get an email from a literary agency, asking if I am represented. This is more like it. One of my goals is to write a book on the year but I haven't had time to look for an agent or publisher. I reply saying 'yes please'.

11:30 The BBC World Service phones. They want to do an interview tonight at eleven o'clock. I'm meant to be asleep by then. However, my passion for the message I want to give out, and my new 'say yes' rule, forces me to … say yes.

11:40 RTE (Ireland) phones. Can I do a live interview at 12.15 pm?

11:41 BBC Five Live rings on the other phone. I ask them to call back in five minutes as I am on the other line.

11:45 BBC Five Live rings again. Yes I can.

12:05 ITV calls and asks for an interview at the café where we will be cooking tomorrow. Yes I can.

12:10 Mobile phone battery dies. I plug it into the mains. I won't be able to do that after tomorrow.

12:15 RTE Radio 1, Ireland's version of the BBC, phones for an interview. Quite a serious one for a change, so sign off very happy indeed.

12:30 An Irish station, i105–i107fm, calls. Asks for an interview at 3.20 pm. I oblige. I'm starting to get tired and stressed. This is too constant, I need some help.

12:40 Another Irish station, Midwest FM, calls. They want an interview there and then. Yes. Questions are the same as most of the other interviews, largely trivial stuff. I'm getting slightly frustrated.

12:55 Friends from Ireland send texts to tell me they have seen me in the paper and now they've heard me on the radio. What on earth am I doing? I forgot to tell them about all this. Oops.

13:15 BBC Five Live phones again. Different show; they want to do an interview plus a live phone-in with listeners' questions at 9.30 pm on Buy Nothing Night. Yeah, why not, I was only meant to be having some home-brew with my friends after an easy day of thirteen hours cooking and hosting.

13:30 Do an interview with Phantom FM on the spot. I should have bought a voice recorder, taped myself answering the questions everyone asks and played it back to them.

13:40 BBC World Service phones. They want another interview for another programme. It's at 5 pm but they need me in the studio, which is about four miles from Claire's house. I say 'no problem' but actually it is. I am stressed. I've agreed to do far too much already because I want to get my message out to as many people as possible. I know I have to do it; I just don't know if I can manage it.

13:55 Get call from *Seoige*, an Irish television lifestyle programme. Want me to come to Ireland to talk about my year and ideas just after Christmas. This could be my ticket across the sea and a chance to talk about the philosophy behind my experiment to the money-loving Irish.

14:00–14:20 Three more small local stations call for quick interviews. Although they don't go to millions of people, I say 'yes' because the presenters are always very kind and supportive and it's a bit more personal.

14:25 Quick lunch. Get email that I am giving an hour's talk at a Permaculture event on Sunday. Nice to be given so much advance notice.

14:45 With Claire, set off in a van to try and find some grains destined for a bin. Go to a local organic food co-op. They give us fifty pounds of polenta, fifty pounds of wheat flakes, seventy pounds of rice and twenty pounds of couscous that were all out of date. Big result.

15:20 Interview with 105–107fm from passenger seat of van, whilst looking for out-of-date food. I'm bored with hearing my voice say the same thing over and over. I try to sound like the words just came to me but I struggle. I've said the same thing in the same manner so many times today. I hope I don't sound unenthusiastic, because I'm not; I am just bored with the same questions.

15:45 Get back to my friend's house, get my bike and cycle to town.

16:15 Quickly print leaflets.

16:30 Buy some food for the last time for a year.

16:45 Buy myself a book while I still can.

16:50 Cycle to the BBC's Bristol studio.

17:15 Do an interview with the World Service. I love World Service interviews – they ask the real questions and waste very little time on trivia. It's a pity I couldn't have done interviews just with them all day. They were the first people not to ask what I was going to miss the most.

17:50 Start cycling home.

18:05 Get a puncture in my back wheel going through a part of

Bristol synonymous with broken beer bottles. The shortest route isn't always the quickest.

18:25 Walk a mile to Fergus's house. I am extremely concerned that I have buckled the back wheel of my bike with about fifty pounds of stuff in panniers. That cannot be good for the tube.

18:30 Fergus and I (in Fergus's van that is now working again), go to Claire's house for dinner. I am going to stay there for my last night of normality. Whatever that means.

19:00 Trying to fix the puncture, I unscrew the rear dérailleur instead of the wheel. I'm dead on my feet and this is the last thing I needed.

19:02 Kick the sofa. Apologise to Claire. Lie on the floor utterly exhausted.

19:03 Watch Fergus attempt to fix the rear dérailleur.

19:10 Tell Fergus it is utterly, utterly hopeless as neither of us has a clue.

19:45 Fergus claims to have fixed it. My body receives a second wind.

19:47 Hug Fergus.

20:00 Still hugging Fergus. He wrestles me off before asking when dinner is.

20:05 I test the bike. Although the front dérailleur is not working, the back gears are fine. I have a functioning bike again.

20:15 I start making dinner. Eat half of it, digest very little, mend the puncture.

21:45 Cycle to town on my half-mangled bike.

22:00 Meet my friends Chris and Suzie. Buy my last beer, and a couple for them, and drink it rather quickly.

22:40 Cycle home to do my last interview of the day at 11 pm.

23:00 Interview with the BBC World Service, the European edition. I remember to speak as slowly as is possible for an Irishman to do, since English isn't the listeners' first language. Not sure it's mine, either.

23:30 Give the last few pennies in my pocket to Claire. Being a student, coupled with the fact she is soon to be the girlfriend of the UK's most financially-challenged man, she gladly accepts.

23:35 Get into bed.

23:36 Asleep.

23:36 Woken by a text message from a producer in an Oscar-winning production company. Not knowing this at the time, I respond with something to the effect of 'yeah no worries mate, I'll give you a shout sometime'. Not my most impressive move.

With my 'relaxing' last day over, it was time to start a year without being able to use money. As I lay in bed, I knew that when I woke up the next day, life would have changed drastically. It was almost impossible to not let that thought weigh heavily on my mind. I had completely underestimated the public and media's interest in a story which hadn't even been written yet and I had a feeling it was going to put a lot of additional pressure on what was going to be a very labour-intensive way of life.

5

THE
FIRST DAY

THE FREECONOMY FEAST

My first day without money.

It felt like it had been coming for an eternity. For weeks, when
people had met me, the only thing they would talk about was my
impending moneyless year. It had become all-consuming. Every
sentence, not just from reporters but also from my friends, seemed
to end in a question mark. 'Why are you doing it? How will you
do it? Will you smell?' I completely understood and expected this
but it didn't make it any easier. Sometimes I just wanted to have
a normal conversation about something other than money or the
lack of it. It was a relief finally to get going.

Knowing how much I had to get done before eight o'clock, I
had set the alarm for 5.30 am. Usually, when it goes off, I let it
ring and vibrate for many minutes before I clamber out of bed.
This time, I was up like a shot, not fired by enthusiasm to start

living without money but in my eagerness to conserve as much battery power as possible. It was good to start as I meant to go on; this year was going to be busy and lie-ins were probably going to be a treat. One of the oddest sensations was not having anything in my pockets, as I had given up using keys weeks before. I had decided never to lock my caravan, as I wanted to start trusting the world more. To be honest, there wasn't much to steal anyway.

The most urgent task of the day was to get the remainder of the veggies we needed for a three-course meal for somewhere between one and two hundred people. I had planned to go with Claire, who had a lot of experience in liberating good food from bins, before the sun came up. But just as we were about to start, I realised that, given the rules I had set myself, I couldn't actually get in the van. In an act full of ridiculousness, Fergus agreed to do my fuel-burning dirty work for me, with Claire. This was great. For the first time in my life, standing up for what I believed in had got me out of some work. Just minutes into my year, the experiment was affecting how I lived. This turned out to be a blessing, as it gave me a few clear moments to contemplate the year ahead. I felt just as excited as I did anxious, but my intuition was that I was going to have a lot of fun.

The day ahead didn't bother me from the point of view of survival, as I would be surrounded by tons of food and lots of people. However, the thought of organising a three-course meal, possibly for over a hundred and fifty guests, without being able either to spend or accept a penny, didn't exactly leave me relaxed. The challenges of the first day weren't related to no-money living, although making five hundred dishes is always a bit easier if you have a few pennies handy. If anything, it was going to be the easiest day of the year, as I would have plenty of food to eat and no spare second to contemplate buying anything. I felt the pressure of making the day come together. I didn't just want to make lots of people a half-decent meal; I wanted to make them the most delicious feast they had ever had. One of my goals was to show that you can thrive, not just survive, without money. If this meal were mediocre, people would be grateful but they wouldn't feel more attracted to freeconomic living and that would be a lost opportunity.

Just as the BBC Breakfast team arrived, Claire and Fergus came back and told me all they'd got was one plum. My heart sank. I'd told the world about this and we had no food. Then they opened the back doors of the van and revealed several hundred pounds of waste vegetables and fruit. The BBC Breakfast team loved this, as it highlighted the very serious issue of the amount of food we waste. They asked us to take it all out of the van to use as a backdrop for the interview. The team decided they wanted two interviews, the second twenty minutes after I was meant to open up the kitchen and meet the volunteers. Money-free for only a couple of hours and I was starting to feel the pressure. I quickly amplified that pressure by telling millions of viewers exactly what I was about to attempt. Even Fergus's sister, who didn't know we were good friends, sent him a message to say she had just heard on the news about some guy in Bristol who was going to be living without money and that she thought he would be interested. There really was no backing out.

The good news was that visitor numbers on the Freeconomy Community website were going through the roof, with four or five new members joining every minute, thanks both to the Breakfast interviews and the fact that Yahoo had put the story on their news homepage. Luckily, a local web developer, Matt Cantillon, who had joined the Freeconomy Community months earlier, had offered to host the website for free. This meant no money was needed to keep the site running, no matter how much traffic it got. This wasn't a one-way transaction; Matt frequently used the site to find free help for the animal rescue project he had set up the previous year and I suppose hosting was the most useful 'skill' he could offer. This summed up the spirit of the community perfectly and Matt and I became good friends during the year.

Fergus, Claire and I got to the venue pretty late and met the first ten volunteers from the Bristol Freeconomy group, ready to start preparing and chopping the food. We got the food together, ready for head chef, Corrine Whitman, to take on a scaled-up version of the BBC television programme, *Ready, Steady, Cook*. On television, the chefs get twenty minutes to create a delicious dish from ingredients they've never seen. Likewise, Corinne had no idea what her ingredients were going to be until she arrived. And she had just six hours to turn a couple of tons of food into a delicious meal for what was to be a packed house of more than one hundred and fifty people.

Although the job of quickly formulating various recipes was extremely challenging at that scale, especially as it was all vegan, she had the luxury of a huge choice. The mix had everything from local vegetables such as rainbow chard, celeriac and kohl rabi to wild mushrooms like chanterelles; from chickweed, nasturtium flowers and rosehips to a plethora of international foods like quinoa, bulgur wheat and couscous, which had travelled thousands of miles from places such as South Africa and New Zealand to end up in Bristolian bins. We ended up with so much food that we had to send some of the volunteers into the 'bear pit', a large circular underpass where drug addicts and homeless people often find shelter, to give out free food for the entire day. Corrine, backed by a growing army of volunteers, managed fantastically well and within an hour had pots and pans of various concoctions on the go, including Fergus's amazing field blewit and wild garlic soup. I had to remind myself not to get too excited; my food wasn't going to be of this standard every day for the next twelve months.

The day went incredibly well. Inspired people unexpectedly volunteered; some of Bristol's best acoustic musicians added an ambience to the occasion and the food was 'to die for'. The punters got free drinks and full service. They couldn't believe their palates and, most kindly, told us so. Andy Hamilton, my self-sufficient-ish friend and home-brewing enthusiast, arrived with one hundred and twenty pints of his best beer, made from locally-foraged yarrow and his allotment-grown hops, as a treat for the volunteers. I finished the last interview of the day, for the *Wall Street Journal*, of all publications, a definite sign that Freeconomy was increasing in popularity. Exhausted and elated, I grabbed a glass of Andy's finest.

Seeing freeconomy work so well in action gave me so much confidence and satisfaction. I decided that if I did manage to make it through the year, I'd end it with something even bigger.

6

THE MONEYLESS ROUTINE

MY FIRST WEEK OF OFFICIAL POVERTY

Even normal change can be destabilising; think how you felt when you had to move house, start a new job or make any changes to your usual lifestyle. You can imagine how it feels to wake up one morning and realise that you can neither receive nor spend a single penny for another 364 days. When I was younger I found giving up chocolate or swearing for the thirty days of Lent a real struggle. Happily, swearing was free and I could continue to do it as much as I liked. My Irish upbringing meant it played an important role in expressing both elation and despair; I had a hunch the coming year was going to contain lots of these emotions.

The morning after the free feast, I woke up at nine o'clock, which was a real lie-in for me. The adrenaline from the last few days had taken its toll; I felt a little fragile and empty. I ate some of the fruit and bread left from the day before and headed for the Permaculture event at which I was speaking. The last two days had been a circus. The real year now began. Instead of being in the newspapers, I would be wiping my rear end with them.

Life without money started smoothly, with no major catastrophes in the first few days. I'd always felt that things would get increasingly difficult the further into my year I went. Stuff would break, I'd run out of supplies and accidents would happen. However, at the beginning, I still had a little bit of everything. This was a good thing. After only a couple of days, I realised that time was my most precious commodity. First, going off-grid was very time-consuming. No switches to turn on energy; even charging up my laptop was a mission. In the dark, I had to hold the wind-up torch in my mouth as I screwed the cable of the laptop's car adapter to the solar panel's charge regulator. It was such a tight space that it frequently took me five minutes to get it plugged in properly.

To compound my lack of time, I spent far too much time in that critical first week speaking to journalists, filming bits and pieces and writing emails to people who had contacted me with questions, opinions and messages of support. This was neither the slow life of self-sufficiency nor the fast life of the city: it was both. Just as the stream started to dwindle, the *Daily Mirror* sent along a reporter for a day, to see how I lived. This was a very positive thing; ten years ago this newspaper would never have been interested in somebody living without money. In some ways it symbolised how far the environmental movement had come, although the credit crunch had doubtless played a part in their decision to follow the story. The article came out fairly well and was mostly positive, if rather sensational. They quoted me as saying 'Gandhi blew my mind, man', when what I had really said was 'I've been inspired by Gandhi in the past'. To convey to the public that I pick nettles every morning for my tea, they wrote,

'Every morning at 7:15 am he crawls on his hands and knees into a field of nettles'. I'm living without money; I'm not crazy!

When you do an interview for the BBC, usually the *Guardian* or *The Times* follows. From my two-page spread in the *Mirror* (with advertisements for Tesco and Boots juxtaposed beneath), I started getting calls from a section of the media I had never engaged with. The *Trisha Goddard Show* wanted me to come on the show with Claire, to ask her how terrible it was to be in a relationship with a man who had no money. I wasn't interested but they didn't stop ringing until I said that conflict really wasn't my thing and that it didn't matter what they asked me, I wasn't going to be negative. They never called again after that.

I also got offers from other publications, including a women's weekly magazine. I checked it out to see what I would be letting myself in for and I was appalled at its stories, from the sensational (a man who tried to kill his wife so he could be with her daughter), to the utterly bizarre. But I felt this was exactly the type of publication I should be talking to. Writing for magazines like *Ethical Consumer* and *Resurgence* is great but there is an element of preaching to the semi-converted. This magazine was certainly not the medium of choice for the alternative class.

Blogs and articles about my experiment popped up all over the world-wide web. A year or two ago, when I first started down this path, I would have been overwhelmed. I would have felt great every time somebody wrote something positive but angry or down every time somebody didn't approve. Now, I didn't care. I'd long ago accepted that the Freeconomy movement was something people were either passionately for or against, with very little in between. I was concerned only with living the way I truly believed in; people could think what they liked.

I do get a bit frustrated when people misquote me or write falsehoods, as once it goes to print or on air it is too late to take it back. One major news channel claimed a friend of mine was paying my National Insurance contribution. I've no idea where they got that from but as far as I am aware, paying someone else's contributions is impossible. I don't think the world of officialdom works like that and I'd be surprised if the reporter didn't know it. Experience made me suspect that they just wanted me to look like a freeloader.

It did feel wonderful finally to start living the way I had wanted to live for so long. There were time pressures and stresses and a lot of physical issues to be resolved but is that not so for any way of life? I believe you become a healthier person – mentally, physically, emotionally and spiritually – the moment you start living the way you believe you should, whatever that may be. Self-discipline is meant to liberate and not constrain the soul. I felt as if a huge weight had finally evaporated from my shoulders.

KEEPING CLEAN WITHOUT TOILETRIES

Soapwort, a natural soap, is not a very common wild plant these days, though you can find it close to streams and in damp woods and hedgerows. However, it is very easy to cultivate, which means you can literally grow your own soap in your back garden.

The frugally-minded can pick up lots of free samples in places where they sell cleaning products but I don't recommend this. Although they are free, they're more environmentally-damaging than buying a whole pack.

You could do what I do and use nothing. When I first tell people this, they usually move back a few feet. I give them the 'armpit test' – a quick sniff under my arm reassures them you don't need soap to be clean. My skin is much healthier since I stopped using soap and since it is no longer dry, I don't have to use moisturisers. I stopped using shower gel long before I stopped using money, because I realised it was very bad for my skin and it made me smell worse, unless I showered every day. The same companies who sell face washes also sell moisturiser; not only do they sell a product that takes moisture and natural oils from your skin, they also sell the one that puts them back in.

If you fancy a haircut, check the windows of local hair salons. Many of them need volunteers for their students and apprentices to practise on. This does involve a certain element of trust!

A TYPICAL DAY IN THE MONEYLESS LIFE

I started to find my rhythm of living without money. By the end of the week I had a nice little routine going. I absolutely adore the morning, so I start the day at five with 'morning oats and oaths'. Oats are a locally-grown food that strengthens me physically, and the oaths are a list of personal ethics and thoughts that strengthen me mentally and put me in the right frame of mind for the day.

Living without money means I no longer go to the gym. Instead, about 5.20 am, I do one hundred and twenty press-ups to warm up and get my blood circulating. Brimming with energy and armed with my wind-up torch, I take off in search of some wild food. In a somewhat mad choice, I'd decided to start my year just at the beginning of the 'hungry gap'; the time of year when there is very little fresh produce available in the vegetable garden. My main wild winter harvests are medlars, ground elder, cow parsley, pine needles for tea, dandelion leaves, stinging nettles and whatever edible mushrooms I can find. 'Jew's ears' are my favourite; a purple to dark-brown, rubbery, ear-shaped fungus. Its texture is fantastic; I call it my vegan meat. They mostly grow on dead elder trees, although you also find them on beech and elm. My caravan is surrounded by dead elders, so I have a fairly constant supply. It's also known as Judas's ear; legend tells that Judas, the apostle who is said to have betrayed Jesus for thirty pieces of silver, hanged himself on an elder tree. I also pick some of the kale and purple-sprouting broccoli that I grow – these aren't wild but are fresh and delicious and essential for getting through the hungry gap.

About six o'clock I head back to my caravan. Living off-grid, I can't just stick the kettle on, so I fire up my rocket stove. Watching the beginnings of the sun coming up over the eastern horizon and listening to the birds tune up, I boil up the nettles and pour the brew into my flask, so I have tea at hand all day. Next comes the rather normal chore of doing the dishes. The not-so-normal part is first breaking the ice in my makeshift outdoor sink. It is really cold at this time of the morning, at this time of year, in the valley where I live. The water is icy but the view is exhilarating.

Before it gets bright I put my compost toilet to good use. My particular model has no toilet seat or bowl, which means I have to squat. This is common in the East, where it's considered the ideal position for clearing the bowels. It's the position we've used to defecate for almost all of human history; our bodies haven't evolved at the same rate as modern toilet-bowl technology (if you can call a toilet bowl a technology). For toilet paper, I get a second use out of one newspaper or another. Using newspapers isn't as bad as it might sound; choosing the right publication is the important thing. I don't find the broadsheets very pleasant, although their content is much better to read. The tabloids work best and at least means they have a useful function. They aren't quite the double-quilted standard I had previously enjoyed but I get used to them remarkably quickly. The best of all, ironically, was *Trade-it* magazine; perfect size and reasonably soft texture.

The funniest moment was when I ripped off a strip of the *Daily Mirror* one morning and, just as I was about to wipe my rear end with it, saw my ugly mug staring right back at me. Needless to say, I carried on; you don't often get the chance to have such utter disrespect for yourself.

Next, teeth. I use a mixture of ground wild fennel seeds and cuttlefish bones (which wash up on British shores from time to time). Cuttlefish bones provide the abrasive needed to clean and get rid of plaque, while fennel seeds both leave your breath smelling incredibly fresh and kill bacteria and everything else which can lead to bad teeth or gums. Fennel is an ingredient in even the most conventional of toothpastes. My toothbrushes come from a batch of seventy or so a friend found in a supermarket bin. They were perfectly fine; it seemed they had been chucked out because their packaging had a bit of water damage. I gratefully accepted them as another potential problem sorted.

I have a quick shave – head and chin – using a cut-throat razor, which I sharpen using the birch polypore (otherwise, aptly, known as the razorstrop) fungus, rather than a leather strop. This is a vegan trick taught to me by Fergus. I finish with a very quick wash under my solar shower. The water is absolutely freezing, as it is winter, but the contraption at least allows me to shower. I fill the black bag from the river for the next day.

It's now seven o'clock and time to boot up my computer. While I'm waiting, which isn't so long because I use Linux, I do another sixty press-ups and ninety lifts above my head with a thirty-pound breeze block. Freeconomy has been growing so quickly over the last two months that I am slightly overwhelmed with work. I dedicate an hour to administering the website and its inevitable queries, then reply to my personal emails. I can't make phone calls so email is now my secondary form of communication, after face-to-face. Once on top of all these responsibilities, I prepare the day's lunch and dinner before starting work on the farm at 8.30.

Work on the farm is extremely varied. One day I grow veggies, another manage the hedgerows, the next (ironically) use my business background to help put together a sustainable business plan for the farm community. I take a break at eleven o'clock, during which time I promote the weekly skill-sharing evening, 'Freeskilling', which I run with the local Freeconomy group. One week it could be on bread or beer-making, the next how to build an earth oven, the next how to make a computer. After another few hours of hard labour, I retire to the caravan for lunch. This is a mix of the food I foraged in the morning, out-of-date bits I garnered from skips the night before and local organic and vegan produce for which I've bartered my skills. While eating, I try to write something: a column, my blog or this book, before getting back on the land.

By 4.30 pm I get the rocket stove on for dinner. I usually cook two days' worth together, to save wood and time. This cooker is extremely efficient, so I eat around five. I devour my meal at a much greater speed than I would like and cycle to the city for a meeting. I tow my trailer; although this adds weight, I can pick up stuff (anything from food to a vegetable steamer) from bins on the way back. The eighteen-mile journey takes about an hour and ten minutes going into the city and about an hour and half on the way back. Going home is all uphill – and I am more tired.

If I don't have a meeting in the evening, I spend thirty minutes chopping up wood, a by-product of our hedge management work on the farm, then get the woodburner going using waste paper and cardboard, some straw, a flint and piece of steel and fine kindling. Once it's going, I get back to the computer for a couple of hours. I do my best to take a quiet stroll through the fields around half-past nine, appreciating the tranquility, beauty and chilly night breeze that surrounds me.

Another hundred press-ups and it's time for a candlelit read. My December reading alternated between Bill McKibben's *Deep Economy*, Henry David Thoreau's *Walden* and Kahlil Gibran's *The Prophet*, a book I have read many times but still learn from. If I don't fall asleep with the candle burned down to the stub and the book on my face, I get up at eleven for a final pee on the compost heap, go back indoors, gaze out at the stars untouched by the city lights and fall into a very nice, deep sleep, recharging my body and mind for the next day's wonderful eighteen hours.

7

A RISKY STRATEGY

Winter can be a difficult time of year for many of us, especially those who live in countries at higher latitudes, like Britain. It's dark when we get up, dark before we leave work and the great outdoors don't seem so great any more. Many people suffer, to varying degrees, from the aptly-named SAD, Seasonal Affective Disorder, also known as 'the winter blues'. And we inevitably spend more of our hard-earned cash, whether on increased energy bills or through the form of escapism known as shopping.

When I announced I was going to start my experiment at the end of November, just coming into the start of the coldest, wettest and darkest part of the year, it persuaded my friends that I had, in fact, finally lost my mind. It wasn't only the weather; there is also very little wild food available between December and March. However, I chose to do a complete year so I could see how it felt to go through four seasons without money. I had to get through winter some time and I thought it best to get it out of the way at the beginning. This was a risky strategy; the first few months were always going to be the most testing and winter certainly wasn't going to make them any easier.

An unwelcome surprise was that it turned out to be – officially – the coldest winter there'd been in my lifetime. I have always loved this season but this was probably because I knew I had the option of a nice warm house with a cooker and central heating when I'd had enough of the elements. There's a reason December is the cheapest time of year to buy a caravan, yurt or converted lorry – no one wants to live outside at that time of year. Not only would I be living in a glorified pig ark, I would be cooking, working, washing, cleaning and emptying my bowels outside in the harshness of a proper British winter.

ENTERTAINMENT

I believe people in countries such as Ireland and the UK drink more alcohol than folk in warmer climates, especially during the winter, largely because many don't think there is much else to do at that time of year. That was how I'd justified spending endless days and nights in the pub. My binge-drinking days came to an abrupt end when I left Ireland in 2002 and became teetotal for a few years, but I still really enjoyed meeting friends for the odd pint on cold wet evenings during the winter. We'd go to the places that had a big fire and have a couple of the finest whilst philosophising, singing or battling it out over a chessboard. Or I might go to the cinema, stick a DVD in the laptop (I've had a self-imposed ban on watching television since 2003, because I had a penchant for frittering away hours watching absolute garbage), listen to some music or call in on a mate.

One of my first realisations was that none of these were now going to be possible, except seeing friends. And even that was going to be extremely difficult, living as I did eighteen miles away from them, with only a bicycle for transport and darkness falling at 4.30 pm. I love my friends but I wasn't going to make a thirty-six-mile round trip in the wind and rain, up and down hills, in the dark, to see them every evening.

I tried to get into the city as often as I could. When I did, I regularly stayed with my friends Cathy, Eric or Francene; all three were really supportive of what I was doing. I'd met Cathy and Eric after they'd contacted me through the Freeconomy website, and Francene was Fergus's ex-girlfriend. Much as I intellectualise that cities are inherently unsustainable models of living and that the pollution and stress that seems to go hand-in-hand with them are really unhealthy, I admit I love Bristol, largely because it's home to some fantastically inspiring people. Many are involved in projects such as Transition Towns, a movement whose *raison d'être* is to build resilient communities by 'transitioning' from our dependency on oil to a more sustainable way of life.

In the first few weeks I had no idea what I was going to do for fun. I'd long since become used to the city way of life, where everything you could possibly imagine was sitting in a shop window waiting for you, price tag attached. I felt isolated living in the country and the public transport was terrible. Not that I could use the bus anyway but it was also difficult to get my friends to come out to my caravan during the winter, as none of them are as fond of cycling as I am. My second realisation sorted out this problem – there wasn't going to be much spare time anyway!

PUNCTURE PROBLEMS

Wanting to meet my friends as often as I could meant I was clocking up the miles on my bike. And because I hadn't established good, efficient waste-collecting routines, I was soon averaging well over sixty miles a week. Whilst much of this was in the country, it seemed that as soon as I got into the city I got a puncture. There's never a good time to get one but at nine o'clock on a cold, wet, winter night, after a physically tough day, it's even less pleasant. Within three weeks, I had used up the few patches I'd had before the start of my year and buying a new puncture repair kit wasn't an option. I tried reinforcing the tyres with old linoleum, so that sharp objects couldn't pierce them, but the little bits that broke off the linoleum served only to compound the problem.

Searching for an alternative, I came across a company, Green Tyres, which made unpuncturable tyres. They used solar energy to power their production and selected their original staff from the long-term unemployed. I really admired the ethos of the company and the fact that their product would prevent a lot of unnecessary resources being used for new inner tubes and repair kits. I wrote a blog about them, so that those who used money could benefit, even if I couldn't. The director of Green Tyres, Sue Marshall, was so thankful she emailed to say she was sending me a few tyres in the post. This wasn't a solution I had considered as a possibility but it was a great reminder that if you trust in life and give without any thought of receiving, whatever you need will come your way when you need it. Which was lucky, as I was only one patch away from walking for a year!

THE 'SLOW' LIFE

Everything, it seemed, took longer. Take washing my clothes. In the past, I'd gather any clothes needing washing, throw them in the machine, take them out when they were done and stick them on the radiator: easy. Not any more. Before I could begin my laundry, I had to make my own soap. First, I hauled waste wood from the city, on the back of the bike, to make a fire. Next, I fired up the rocket stove to boil some water, into which I stuck some

BOOKS AND PAPER FOR FREE

Reading and writing are two of my favourite ways to spend time, especially in front of the woodburner in winter when the wind and rain are pounding against my caravan. Thankfully, you don't need cash for either.

For books, the library is your obvious bet. Those in rural areas might find a mobile library visits. However, not everyone finds the library ideal. You have to hand the book back within a certain time or incur a fine and not everyone can read it in the allotted time. And the library may not stock the book you want (though you can ask them to order it), especially in small towns and villages.

Websites such as ReaditSwapit (www.readitswapit.co.uk) and BookHopper (www.bookhopper.com) allow you to swap books you no longer want for books you would like to read.

I've also organised book-swapping evenings; an offline version of the websites, with the benefit of being much more personal. You can get rid of the books you don't want, get ones you do and meet like-minded people all at the same time! If you want something completely different, take a look at Book Crossing (www.bookcrossing.com) – I'll let you check out this little gem for yourself!

For writing paper, I use old till receipts from a shop in the city; these are great for leaving notes and would otherwise be binned. You can also make perfectly good paper and ink from mushrooms.

soapnuts (*Sapindus mukkorossi*, a plant native to Nepal), 'foraged' from a local eco-store that had gone out of business. I boiled the nuts up for about half an hour – constantly feeding the rocket stove with old broken-up vegetable boxes – and lo! I had myself some detergent. This was no ordinary detergent; not only did it clean just as well as supermarket brands, it was much more environmentally friendly and certainly not tested on animals. With no way of heating lots of water, I'd put the clothes and detergent in a my small, makeshift sink with some icy-cold water, scrub for forty minutes, rinse for twenty minutes and hand-wring as much water out as I could before hanging them up to dry. In winter, clothes can take days, if not the entire week, to dry outside.

Not only washing clothes took extra time; everything did. Making a cup of tea took about twenty minutes. I decided it was sometimes more pleasurable to just not drink tea. Going to the toilet was equally time-consuming. First, I had to make sure the coast was clear; I didn't want to upset local people who might arrive on the public footpath near my compost loo just as the belt of my trousers hit my ankles. Then, my hole in the ground inevitably seemed to be full at exactly the wrong time and I'd have to spend ten minutes, with tightly-clenched buttocks, digging a hole half the size of my leg while praying I wouldn't have an accident and have to re-start the whole clothes washing process.

When it got cold, I couldn't turn the central heating on. There'd be wood-chopping, kindling-gathering, paper-finding and fire-starting before the fire even got going. Then it took a further thirty minutes to warm the caravan. There isn't, unfortunately, a timer on a woodburner. It all sounds like a nightmare but it's wrong of me to portray it like that. There are a lot of environmental benefits in this way of life, which I believe outweigh the inconveniences:

Time to wash clothes with money: 10 minutes. Time to wash clothes without money: 2 hours 15 minutes.
Water used to wash clothes in a machine: 100 litres. Water used to wash clothes by hand: 12 litres.

Water used in a flush toilet each day per person (according to American Water Works Association Research): 70 litres. Water used in a composting toilet each day per person: 0 litres.

If the people of the UK successfully made the transition to compost toilets, not only would we save two billion litres of fresh water per day (figures from Waterwise), we'd also have a lot of great compost, to put back into the soil that which we've taken out.

Average household energy bill during winter: £400 (more than my entire sustainable home cost!). My average monthly energy bill: £0. (This difference is like a person on the minimum wage having two weeks off work during winter.)

I discovered that I had no such thing as a work life–social life–private life balance. I just had life. Instead of doing an evening course paid for by the money I earned in a normal job, my learning came from being out in nature. I became acquainted with the sounds of local birds and learned more about squirrels through observing them than I ever could on the world-wide web. I realised that Jew's ears mushrooms have a fondness for elder trees and that there's a big difference between burning elder wood and alder wood.

My favourite times were when it rained heavily. I'd listen to the rain crashing on the roof with a real appreciation for the shelter that was keeping me dry and protected and for the tree that supplied me with the wood that was now keeping me warm against the wind. Not to mention my thankfulness to the guy who made the woodburner. Such gratitude increases as you get closer to nature and the things that you use; the more degrees of separation you have, the less you appreciate them.

Because of what I was doing and the exposure it got, I did a lot of writing. I'd dreamt of living with nature for years; years when I had complained that I could never find the right space to think, read and write. Sitting in front of the woodburner, watching the embers glow and looking at the moonlight filtering through the trees, was perfect. My thoughts were clearer and I wrote articles in half the time it would have taken in the city.

It wasn't all nature and coping with my, perhaps inevitable, feelings of isolation. There were free film nights in the city and most weeks I'd go to Freeskilling. These evenings were so much fun and very informative, and gave me a real sense of doing something communally. They were also a great way for local people, who couldn't afford to pay £10 or more for a workshop, to learn the traditional skills they would need for a sustainable future. Through Freeskilling, I got to meet loads of new friends every week and learned new skills at the same time. After each session, we would often go back to somebody's house, rambling into the early hours about what we had learned and how we wanted to put it into practice. I organised the evenings with two local Freeconomists, Lucy and Amanda, with whom I became good friends very quickly. Whilst neither had any inclination to live without money completely, both were passionate about skill-sharing and the need to rebuild our crumbling communities through sharing resources. Their enthusiasm and energy was a great source of inspiration.

Living the slow life is definitely more time-consuming but I'd rather have it consumed this way than in watching a reality television show in the room we call 'living'. If we want to be truly sustainable in the long term, I really believe that this is what we need to do. The modern conveniences we have grown to love, the washing machines, dishwashers and cars, come from an industrialised society, with the pollution and environmental destruction that go hand-in-hand with it. If I didn't really believe this, I wouldn't put myself to so much trouble.

My only frustration, I suppose, was that people around me hadn't, understandably, really grown to appreciate how much more demanding this life was both of my energy and my time. They expected me to live the fast life alongside my slow life, go to meetings in the city two or three times a week and do all the things I had to do. Sometimes I wished they could swap places with me just for a couple of days. But I had made my bed; there was no point moaning about the state of the sheets.

8

CHRISTMAS WITHOUT MONEY

Christmas began as the celebration of the birth of Jesus of Nazareth, a man who, by historical accounts, spent the last years of his life preaching simplicity. And to some people it still is. However, for the majority of people in western society, what it has become is far removed from what it was originally. The festive period has now become the most important shopping time of the year for most retailers. According to Deloitte, in 2008 the average Briton spent £655 on gifts, socialising and food. That's over £36 billion for the nation – and 39% of it was on credit. UN figures show that during the twelve days of Christmas 2008, world-wide, 207,360 kids (the equivalent of the population of a small city) died of starvation.

What was traditionally a time to relax with family and friends has gradually become a huge source of stress for many people. According to the *Daily Telegraph*, 8th January is the busiest day of the year for divorce lawyers. Mind, the UK's leading mental health charity, says 25% of people suffer from depression soon after Christmas, most of it due to the financial hangover of its aftermath. It's an expensive party in more ways than one.

Given the amount of pressure, subtly applied through huge advertising campaigns, to buy the biggest and best Christmas presents for your family and friends, many people questioned the rationale of becoming moneyless at the exact point of the year when it seems everyone else can't spend enough. It did feel a bit strange, but to be honest, not buying presents didn't really bother me; my adult friends knew what I was up to and weren't expecting anything. And I think the fact I demanded they buy me nothing in their turn reassured them I wasn't playing Scrooge. I was concerned about my nephews but I decided it would be a good chance to explain to them why Uncle Mark wasn't as kind and generous as Santa Claus (who for a guy that travels the world using a reindeer and sledge sure has one hell of a carbon footprint).

What did bother me was the thought of not getting home to spend Christmas with the people I love. The family had gone to relatives' houses almost every year I could remember and my folks loved having us together for Christmas. Did leading what I deemed to be an ethical life mean I would have to upset my mother?

My main hope stemmed from a phone call I'd received from a television programme in Ireland on the first mad day of my year. They wanted me to come and be interviewed on a daytime lifestyle programme, *Seoige*, presented by Gráinne Seoige, voted Ireland's sexiest woman. This was upmarket bartering. Guests were usually paid quite handsomely for their ten minutes of work but this wasn't an option for me, so I politely declined. They also offered me return flights and trains and buses in between. In 2006, I'd vowed to never fly again, so I told them that if it did happen, I would only go by ferry. And as one of my principles was to accept only as much as I needed in life and no more, I felt that even a train or bus ticket would have been too much.

For most of December, *Seoige* was merely a possibility. My other idea was to cycle to Fishguard (my nearest ferry terminal for Ireland), hitch a ride with a lorry driver and cycle all the way to the north-west. But few lorry drivers take hitchhikers now, because of insurance complications. The more I contemplated this option, the harder it seemed. Doing it with money would be hard enough; without money it would be perilous at this time of year.

Christmas approached and I had no concrete solution to getting home to my folks. I was increasingly frustrated with the limits that living this way was putting on me. But just when I thought it wasn't going to happen, I got a call from RTE to say they wanted me on the show and would email me return tickets. This was the solution to my only real obstacle – getting across the Irish Sea. Everything else I was fairly confident I could pull off without money but the Irish Sea is a long swim. I decided to hitch the whole road trip, from Bristol to the north-west of Ireland. If I was lucky, it would take me two days; if not, Christmas Day would probably see me walking up a deserted main road, without food or shelter. I wouldn't even have a working phone; without credit, I couldn't receive calls outside the UK.

I didn't have a lot of time to prepare. Food was the main issue; I decided I'd better collect and make enough for three days. You never know how long hitchhiking is going to take – sometimes you can wait hours (though that has never happened to me). It was only about a four-hour journey to Fishguard but I knew I needed to get there before sunset, which would be at about 4.30 pm. Getting a lift in the dark is difficult and sometimes a bit dangerous, depending on what type of road you're dropped off. Sometimes you can get dropped off at the wrong end of a city.

That was my worst case; it would have meant walking for miles, with a heavy backpack and without a map, to get to the next hitching point.

I started in good time on 23rd December, leaving at 10.30 in the morning to catch the ferry from Fishguard at two o'clock the following morning. Starting my journey the day before Christmas Eve was an interesting experiment. Before setting off, I wondered whether the Christmas spirit of old would prevail. Would everyone who saw me want to help me on my way or would they be too stressed and busy even to a see a hitchhiker on the side of the road?

My hitchhiking experience tells me that you have to get into the right head space. When your body language portrays confidence, openness, optimism and jolliness, getting a lift seems like child's play. When you're a bit off-weather, it feels like no one wants to know. I got into the groove, got smiling and got out on the road. This wasn't hard, as I love the adventure of hitchhiking. On the bus or train you know that you'll get on at A and get off at B and rarely speak to anyone in between; hitchhiking, you never know what is going to happen. If you can get over the uncertainty, it really puts the excitement back into travel.

TIPS FOR HITCHHIKING

Location, location, location. A great spot makes the difference between waiting five minutes and waiting two hours. Find a place where you can be seen easily, where the traffic is moving at less than 45 miles an hour and where a car has enough time and space to pull in safely. No driver will risk their life to pick you up.

Look happy. Few people want to share their car with someone who looks miserable! Smile and be friendly.

Wear some bright clothing. It also helps if you look clean and approachable. Have clothes to match all sorts of weather.

Keep luggage to a minimum.

Know your route. Know the roads you want to take and avoid motorways; it's illegal and difficult to hitchhike on them in most countries. Some people like to use a whiteboard and marker to show their destination but I don't bother.

Trust your instincts. If you have a nervous feeling about getting into a car with someone, make a polite excuse and don't. But don't be too fearful – I've hitchhiked since I was a kid and never once had a problem, although there are issues to consider regarding your sex.

Don't feel down. Don't let cars going past get you down and don't criticise the drivers who do. A positive attitude is vital to get you where you want to go!

In many respects, hitchhiking is a good metaphor for life!

My positive mental attitude seemed to work and I made it to Fishguard in less than five hours, not much longer than if I'd driven myself. However, this did leave me with about twelve hours to pass in an empty ferry terminal. As I sat there, alone, I wondered how busy the airport was and what impact cheap flights have had on ferries. The good news was that I had half a day to read in silence, something I crave. The bad news was that it was very cold. There was a television room, in which the heating automatically went on as you walked through the door. But I was the only one there, and my self-imposed rules said I couldn't go in, because the heaters would be on just for me. I spent the entire day glancing at the door, knowing only too well that if I went in my body could thaw out. Moments like this made me wonder if I was taking things too far; then I visualised the sea-levels rising over low-lying countries such as the Maldives and went back to reading my book.

A terminal attendant came to tell me about this warm room and even went in to turn the television on. When I stayed where I was and he asked why, I didn't know what to say. If I told him the real reason – that I wasn't going in because of climate change – would he have thought me insane or would he have respected my views? Arrogantly, I didn't give him the chance, but muttered that I was really comfortable where I was, though I appreciated his offer. He left, looking at me curiously. Around midnight, another passenger arrived and headed straight for this room of luxury. The sound of ten electric heaters turning on and my footsteps aiming in that direction quickly followed each other; I convinced myself that if the heating was going on anyway, I may as well make the most of it.

I got about thirty minutes' sleep on the three-and-a-half-hour journey to Rosslare. On the ferry, I had some problems with drinking water. I'd assumed I would be able to get some water either from the bar or in the toilets, so hadn't bothered filling up my bottle. That was a false assumption; one of the restaurant cooks warned me that the tap water was far from drinkable, pumped full of chemicals to make it clean. I'd lived on a boat, so I should have remembered this. Already slightly dehydrated, I had to start hitching without water, knowing that it could be at least five hours before I could find somewhere I could refill my bottle.

It was Christmas Eve and I had about twelve hours to get from the most south-easterly point of Ireland to the very north-western coast; roughly three hundred miles. On the motorway, the route most long-distance drivers would use, this would normally take about six and a half hours. The problem was I couldn't risk taking the motorway route. It's illegal to hitchhike on a motorway; you can't get dropped off on them and the slip roads are always hit and miss. So I had to go on the smaller roads, which meant I would have to get a lot of small lifts.

I hightailed it out of the ferry port to try to get ahead of the cars and ended up running for over a mile to get to a good spot. Rosslare is a very quiet town – whenever the cars get off the ferries and leave, nothing much else passes through. My ferry was the last one before Christmas; if I missed the traffic coming off it, I was in trouble. But my luck was in; a lorry driver took me a few miles up the road to a great location and I was off. And that luck continued all day – the longest I waited for any lift was about ten minutes.

By 3.30 pm I was knocking on the door of my folks' house, to the complete surprise of my mum, who thought I hadn't a hope of making it the whole way before Christmas. It had taken me just under nine hours to get from Rosslare, about the same as if I'd had my own car and taken a couple of breaks. Humanity, it seemed, has a really positive side that we don't hear about too often. Between Bristol and Donegal, Ireland I had fifteen hitches altogether. Compared to a cheap flight, the results were varied.

	Time	Cost	Adventure	Comfort
Bristol – Donegal (with money)	8 hours	£55 (flight) £25 (bus)	Low	High
Bristol – Donegal (without money)	29 hours	£0	High	Low

This table tells a story in itself – have we as a species swapped adventure for convenience?

It was really fascinating to see the types of people who gave me a lift. Every one of them had an average car, which made me question whether the more wealth you accumulated, the less you wanted to share. Most of the people who gave me a lift said that they had hitchhiked when they were younger, so there was a definite empathy. From their stories, I could tell that some of them wished they were back on the road and feeling the adventure of hitchhiking again; in some cases it almost felt like owning a car was something inflicted on them. Although there was only one person in the car every time, they were a very diverse group. Contrary to popular opinion, the majority were women (about three out of every four lifts I receive are from women). One, who had just come off the night shift and faced a fifty-mile drive home, told me she always picked up hitchhikers whenever she saw them, just to keep herself awake, and in ten years had never once had a problem.

One incident, which really uplifted me, came after I had got out of a car only to realise I'd left my freshly-filled water bottle behind. This was my only water vessel and I hadn't drunk much for twelve hours. Without it, I would have had to search in a bin for an old plastic one and fill it in a toilet somewhere. One hour, and another hitch later, the guy in whose car I'd left the water bottle pulled up. He'd spent forty minutes searching for me so he could give me my bottle back. I'd told him I was living without money and he guessed the bottle was pretty important to me. This guy had told me he had just spent two years in prison after a fight outside a nightclub. And here he was, going to all lengths to make sure a complete stranger had his water bottle. This reinforced my belief that there is no such thing as a 'good' or 'bad' person; each of us is just as capable of huge acts of kindness and generosity as we are of causing harm. Our challenge, as evolving humans, is to maximise the former and minimise the latter.

Another guy had heard me on the radio a few weeks earlier as he drove along the exact same stretch of road where he'd picked me up. He was fascinated by it all and asked me to come and stay in his place in Waterford, on the south coast, to help him build his new house. I promised that if I got another chance to come to Ireland within the year, I would take him up on the offer. Everyone I got a lift with was extremely interesting in their own way, each with a story to tell and a great knowledge of their local area. In almost every case, we parted having learned something from each other.

A CASHLESS CHRISTMAS ...

Now I was home, it was time to think about how to spend an entire Christmas without buying anything at all. My friends like a pint at the quietest times of the year but at Christmas they move into top gear. This Christmas had an extra edge; my mate Barry was getting married and his stag party had been arranged for 27th December. This meant – obviously – a huge traditional Irish drinking session was on its way.

In my days with money I was, like almost every other Irishman, one of the first at the bar buying a round of stout for the crew. Given that it was a stag night, my Irish instinct was telling me to get everyone a pint of the best and a double tequila. You can imagine my discomfort at having to go to the bar knowing that I couldn't even buy myself a drink, let alone the stag. The guys were great; only I felt a bit off about it all. They tried to fill me with booze, though I repeatedly refused, trying to make a point about the year not being about freeloading. It was a wasted effort. Before I knew it Marty, my best friend since I was about six years old, had put three pints of organic cider in front of me and told me I could return the favour by giving him a plug on the Gráinne Seoige show. I was bartering pints for street cred.

My awkwardness increased as the night went on. After the pub, the lads said they would pay my taxi fare and the stag said he would pay for me at the nightclub. This I had to draw the line at, but I couldn't win. I didn't want to be freeloading on my friends but I wanted to go and celebrate with my mate. In hindsight, I think I took the cowardly option. I went home, putting my desire not to be seen as a freeloader over spending one last night with my mate as a free man.

This wasn't the first time I'd been in such an awkward social situation; I'd already had a few in Bristol. Whenever I went out with acquaintances, they would start the conversation with 'am I allowed to buy you a drink?' When I answered 'no', they kept trying and if I finally said 'yes', they'd say 'oh yeah, you won't buy one but you'll let me buy one for you!', and I'd say 'no thanks' again. There were few exceptions and although it was always done in jest, my over-active male ego didn't really enjoy it. The stag party was perhaps the most extreme occasion and the one time I felt I'd made the wrong decision.

Waking up on Christmas Day was odd. I'd been a really good boy all year and was hoping Santa had brought me the latest video games machine, complete with solar panels. But I awoke to an empty stocking. It was completely refreshing; in the past we'd often get each other the most needless and uninteresting things you could imagine, faking our excitement as we unwrapped the layers of wrapping on another pack of socks or an electric foot massager.

All my family are Catholic; my uncle is a priest who does great work in the community, so we always say grace before everyone gets stuck in. This is a practice I love, for no other reason than it makes everyone think about where their food comes from. While everyone else had the normal festive feast (turkey, beef and roast potatoes for the main course, swiftly followed by jelly, pudding, custard and cake), I ate my own little, more humble, stash. This was pretty much the same as the dinners I had eaten for the previous four weeks; mainly food I'd brought with me, along with some steamed root vegetables my mum and dad had got from a local organic farmer. There were plenty of brussels sprouts, so I was more than happy.

I am very lucky to have an understanding family. My relatives bent over backwards to accommodate me, even though I didn't really need much accommodating. To a lot of families, I would have been the awkward son, always creating hassle, but I was surrounded by loving, supportive people. The great thing about this Christmas was that I got to spend a lot of time with my folks and we had a good time together. In years gone by, I would have been spending a lot of money getting myself a hangover or out at the January sales, doing the stuff you do when you have some disposable income. Having no money forced me to do the simple things. We spent two or three hours every day walking the coast, playing beach tennis or going for a wander through the woods. Other times, we'd sit and talk together or play cards. This was normal in the Ireland of thirty years ago but is becoming more alien to a country bitten by the Celtic Tiger.

Having a wash in Ireland was challenging. I'd left my solar shower at home but to be honest I didn't really care. It wasn't very useful in the winter but it did give me a way to sprinkle water over myself, even if it was icy cold. Over here, it looked as if my best option was the Atlantic Ocean but given that it was one of the coldest Christmases in living memory, it wasn't something I wanted to do every day.

For the first week, I just didn't wash. Then, because it was coming into a new year, the time of fresh starts, I decided I should polish up my act and so off to the beach I went. It was freezing cold, as you expect at Christmas in Ireland. Getting in was harder than being in. I had to do some exercises first to get warm enough to strip off before sprinting to the sea, where I knew the best course was just to dive straight in. This was easier said than done. The water was up to my rear end before I took the leap. But it was surprisingly good and much more invigorating than a hot shower. The water felt amazingly clean on my skin and the sun was shining down from a blue sky, doing its best to negate the chilly westerly breeze. The surrounding green hills and mountains converged to the beach. I could not imagine a more picturesque, beautiful bathtub. It was cold and it wasn't very convenient but the setting, and the feeling of being with nature, more than made up for it. I think we have swapped the experience of being exposed to the elements for comfort. We have, in the words of Roger Waters of Pink Floyd, become 'comfortably numb'.

Not buying something, regardless of how much healthier it is for your body, is not very good for the economy. You'll never see a magazine advertise this approach; you'll see a model you could one day look like, if you only buy the product that they hold between their palms. Years of multi-million-pound-backed propaganda is hard to delete from people's minds. When I told people I only washed once a week in winter, without soap, they did the thing where they scrunched their faces, said 'ooohhh' and asked me 'do you not feel dirty and smelly?' I'd explain how soap was completely unnecessary but it would fall on shocked ears.

My other piece of advice, if you don't want to use soap or wash so often, is to eat organically-produced, fresh, vegan food. Sweat is little more than salty water if you are healthy; if you put trash in your body, you must expect to come out smelling like it. Since giving up both meat and dairy foods (both especially bad at causing this effect), I've found a massive difference in how I naturally smell. Avoid or reduce both if you want to start going without soap. Being vegan has also meant that I don't need to wash my dishes with detergent, as washing-up liquid is only necessary when you are cleaning dishes that are likely to have bacteria such as salmonella and campylobacter on them. According to the UK's Food Standards Agency, the rise of infections of these bacteria is due, in part, to the terrible conditions in which we both house and kill animals.

LOW-IMPACT TRANSPORT

Transport is no longer seen as a complete luxury. We rely on it to get to work, to see family and friends dispersed around the world, and to eat. Transport makes up 21% of the UK's carbon dioxide emissions, so it is important that we come up with solutions – and quickly – if we are going to prevent serious climate chaos.

Some of these solutions are already available. Organisations such as Liftshare (www.liftshare.com) and Carshare (www.carshare.com) enable people who are taking the same journey to travel together. It's like hitchhiking, organised online, making it safer and less uncertain.

Other projects, such as the City Car Club, are also helping. This is a 'pay as you go' system, which makes driving much cheaper and cuts down on numbers of cars produced, as several people can share one, using it only when they need. And if you offer your journey up as a lift-share, you can help the environment even more.

Hitchhiking is becoming a thing of the past, which I think is very sad. Once in a generation perhaps, someone is killed while hitchhiking, the media sensationalise it and no one hitchhikes for a long time. Hitchhiking is a great adventure, you meet amazing people with lots of local knowledge and you sometimes decide to go to places you had never intended. My favourite journeys always involved sticking my thumb out.

Walking and cycling, I find, are the most relaxing forms of transport. They're really natural exercise and save your gym fee into the bargain. I have friends who drive to the gym, get on a bicycle machine, do forty-five minutes and then drive home! I tell them they should save themselves the gym fee, the cost of the fuel, car tax and insurance and cycle to the gym and back without going in!

Two organisations which have made walking and cycling a lot more fun, safer and more enjoyable are The Ramblers Association (www.ramblers.org.uk) and SUSTRANS (www.sustrans.org.uk)

NEW YEAR'S EVE

Consuming as much alcohol as humanly possible is synonymous with 31st December in most of the western world. Indeed, in Ireland, they do not limit it to what is humanly possible.

Until now, New Year's Eve had gone something like this. Wake up, eat a quick breakfast, phone mates, get to the pub, convince the landlord I am not an undercover policeman and start drinking by ten o'clock. That, however, was when I had money. This year, it was going to have to be different. Even the crap pubs seemed to charge an entry fee on New Year's Eve and tickets for the lowliest of nightclubs started at £20, with drinks at a huge premium. This was irrelevant to me; I couldn't afford to look at the barman, let alone ask him for a drink. Even my parents were partying. My mates went out as normal but to save myself the mental turmoil of the stag party, I stayed in. As 2008 became no more, I lay in bed writing the start of this book.

This was a blessing in disguise. For once, I began a new year without feeling like someone had drained every millilitre of water from my body, gripped my head in a clamp and repeatedly smacked me on the back of the skull with a rubber mallet. I hit a deserted beach for a stunning early morning walk with my folks, looking forward to the year ahead instead of wishing somebody would use a rusty saw to disconnect my head from the rest of my body. This, I decided, was how I was going to spend New Year's Eve from now on, money or no money.

Normally, on New Year's Day, I would arise and write a list the length of my leg of things I resolved to do (or not do) in the coming year. But what on earth was I going to give up this year that I hadn't already? There wasn't much left. Food? Water? Oxygen? Hope? To retain the last, I decided to call off the resolutions: enough was, finally, enough.

RETURNING TO AN ICE BOX

Christmas was over before I knew it and I had to make my way back to Bristol. This time, the journey had to be broken, to complete my end of the bargain by talking about my experiences so far on *Seoige*. And a bargain it was for them – I couldn't accept their taxi rides or even their food; it wasn't vegan, let alone organic or local.

The interview went well, though I sensed Gráinne, the presenter, wasn't my biggest supporter. This was fair enough and I couldn't blame her. She'd spent half her life climbing the television ladder to get to the point where she could earn a lot of money. It could have seemed I was saying hers was an unethical way to live. After the pleasantries and the 'tough' questions (that I'd heard a million times), Gráinne attempted a question from the left field. 'You've been quoted as saying that if you have £1,000 in the bank and a child in Eritrea dies from starvation, in a way you have some responsibility for that child's death. Should you not be earning money and giving it to charities in developing countries?' Gráinne asked with a slight grin. 'Earning money from, and supporting, a system that keeps these people in poverty in the first place and then gives them some of the profits in the form of "strings-attached" aid or World Bank and IMF loans is no more ridiculous than Shell or Esso giving Greenpeace or Friends of the Earth £10,000 to help clear up the destruction that they inevitably cause. Would it not be better not to cause the destruction in the first place?' I replied, quickly followed by 'But yes, if you insist on earning money and, collectively as a nation, riding on the backs of the less fortunate, you should give as much as you can to charities'.

Seconds after I mentioned the names of two of the world's major oil companies, both of which advertise on RTE, I sensed Gráinne was receiving instructions through her earpiece from the producer. Suddenly the interview was at an end. My intuition told me that they didn't appreciate me implying two of their biggest funders behaved in untoward ways and probably concerned I was getting too political for a nice Tuesday afternoon lifestyle programme.

After the interview, it was back to Rosslare to get the ferry home. If I thought I'd got lucky with lifts on the way over, the way back was even easier. I sprinted off the ferry to get ahead of the traffic in Fishguard, stuck my thumb out at a place where I wouldn't normally have bothered and within two minutes I had a ride with a lorry driver on his way to Germany. Not only was he going in my direction, he was going within a five-minute walk of my destination! Part of me was disappointed, as it meant that the adventure was coming to an end and I would miss meeting some new people. The other half was delighted; it had been a long journey and I would definitely be in a warm bed for the night.

I thought the hardest part of my winter – an overseas holiday without any money – was over, but I returned to weeks of snow and ice. In the city, snow softens the harsh industrial edges and makes everyone feel they are living closer to nature; in the country it covers the hills and valleys with colossal white blankets. I love snow but it did make my life a lot more difficult. For two weeks, the small country roads were covered in snow or ice as the local council hadn't enough grit to go around. Driving in these conditions can be treacherous enough; when cycling, it's extremely dangerous. But to eat and to get waste wood, I was very dependent on my bike, unless I wanted to spend the entire day walking.

I ran down my reserves within a couple of days and had to look for new solutions. For wood, my first thought was to chop up the pallet that was the base of my front doorstep. But then I stopped and thought about what I was doing: contemplating burning part of my house to stay warm for a few days. This, I thought, was exactly what humanity was doing; consuming its assets for very short-term objectives, many of which are a lot less necessary than keeping warm. The doorstep stayed and I cycled off to get supplies. A couple of times I cycled for seven or eight miles on solid lumpy ice, on which, I soon realised, it is incredibly unpleasant to fall on your ass.

Apart from everything else, it was just bloody freezing. Most days it didn't rise above 0°C and on many nights it fell to -6°C; the temperatures felt even lower in my valley. I lived in a tin can; with the woodburner on, it was fine but sometimes I didn't get home until late and just wanted to get into bed and sleep, so there was no point in lighting the burner. It was always freezing in the mornings. Sometimes the outside of my duvet was stiff when I woke up; the insulation in my caravan was so poor that even if I lit a fire in the evening, it would be cold within three or four hours of the last log. This wasn't really serious but made it very difficult to get up at five o'clock in the morning.

9

THE
HUNGRY GAP

In a world of cheap energy, highly efficient logistics and vacuum packs, it's summer all year round for your diet. Even in the shortest days of winter, grapefruit, pineapples and tomatoes can come from almost any corner of the planet within days. But before the technological developments of the eighteenth century, the vast majority of a British meal came from somewhere in the UK. Only life's treats – the sugar and spice – came from further afield. Between January and March, food was scarcer than during the summer, as the crops growing at home were limited and few could afford to buy the bulk of their food from abroad.

Living without money meant I returned to the diet of the England of the 1700s. You can grow enough to keep yourself going between January and March but it means eating the same things most days. On a locally-grown diet, you are limited to root vegetables and crops like potatoes and barley grain for the base of the meal. I think of barley grain as the 'English rice', yet it's a grain very few people use, despite it being both delicious and nutritious. While eating locally during the winter seemed daunting, part of me was excited. There is something about the flavour of food you've grown or picked yourself that no spice in the world can match. Contrary to even my expectations, I quickly came to really enjoy my evening's repast, eating each steamed vegetable individually, to savour both its taste and the taste of the British winter.

I hadn't accounted for the heavy rain that came between December and February. At the farm, down by the river, were some polytunnels; large, cheap greenhouses used to grow foods that need a slightly warmer climate than the UK can provide. I am a bit torn about polytunnels. They are made of plastic, with the embodied energy, pollution and suffering that goes with it, yet they allow us to grow food throughout the year, meaning we have to import less, so using much less fossil fuel. Without them, sustaining more than sixty million people, all year round, is unrealistic, at least in the short term. These modern greenhouses were a great source of nutritious, fresh food for me during the winter, until we had two days of extremely heavy rain, accompanied by inevitable flash flooding that filled the polytunnels with about three feet of river water. The flood was fine in itself; no major damage was done. But the river had been polluted, in various ways, for several years. Now, not only was I unable to drink the river water, I could no longer safely eat the vegetables I had spent months preparing, planting and weeding.

Throughout much of the world, including the UK, we have an unnatural system; when water comes from a tap, few people really care about polluting a river. As far as most people are concerned, it'll get cleaned up before they have to drink it. Floods are natural events. While it's impossible to say it has increased because of climate change, since 2004, flooding in the UK has increased both in numbers and severity. Dr Tim Osborn, a leading expert on flood risk due to climate change, estimates the chances of three or more days of heavy rainfall have doubled since the 1960s. I suppose it is common sense to think that the more you disrespect the planet, the more extreme the consequences will be.

This flood caused me no end of problems. Instead of eating the food I had grown, I had only a few vegetables left in another field. Thankfully, one of these was kale, a sturdy, robust crop, essential for anyone with ambitions to live completely on local food for a year. It is very nutritious and grows through the hungry gap. The loss of my other crops meant I was going to have to find alternative sources of food, which would mean more time and more cycling. I had to eat a bit more waste food than I had planned and do some more bartering. For me, it was important to do a variety of work when bartering, and with a lot of different people, not just 'alternative' environmental types. One day I worked with a Hungarian man, Peter Horvarth, who

WILD FOOD FORAGING

Foraging for food, whether in the wilds or an urban neighbourhood, can be done by anybody. However, I recommend that you follow some guidance to begin with and take reasonable care at all times, as some wild plants can be toxic. To get started, I'd recommend:

A little book called *Food for Free* by Richard Mabey; get it from book swapping websites such as ReaditSwapit. co.uk

Taking a wild food foraging course – you'll find an excellent one at wildmanwildfood.com

Check forums such as Selfsufficientish.com for hints and tips supplies snacks such as bhajis and pakoras to Bristol's food shops. For five hours' work, he gave me more than thirty falafels, which had to be eaten within a week. Whilst the quantities involved were not entirely my idea of healthy eating, I think folk in pre-industrial times would have been very grateful for such a bounty at this time of year. I also did a bit of casual work in a Health Food Co-op in the city, which I felt was really important. I wanted to include both city and country folk and to prove you can do this no matter where you live.

THE ENERGY GAP

Never having lived off-grid, and coming from a fairly normal background, I'd got as used as anyone to seemingly infinite energy being available at the touch of a button. Spending an entire winter, the time of the year when daylight is most limited, using only solar power, was an interesting – and often frustrating – experience.

It gave me a new appreciation for energy; it was no longer endless. The interest from the media at the start of the year meant I'd taken on a lot of writing for magazines and newspapers. This put a huge strain on my battery; it regularly ran down. I found this quite frustrating at some times and downright infuriating at others. Learning that I couldn't have all the energy I wanted when I wanted it was a real test, as was learning that if I did, I had to find a way of producing it.

One solution I found was first to write my articles using pen and paper, then type them on to my laptop, to save the solar energy running down as I structured my thoughts. However, I couldn't buy pens or paper, so I needed a solution for that as well. I had two options. The first – the ecological but time-consuming option – was to make ink and paper from mushrooms; I'd learned how from Fergus. (My biggest advice to anyone considering living without money is to befriend Fergus; the extent of his knowledge is equalled only by his incredible willingness to share it.) But given how much writing I had to do, I had neither the time nor the natural resources to use this method very often. Instead I turned to waste.

Paper was easy; I took sheets of A4 paper from paper recycling bins; almost always only one side had been printed. I gave it another use before it went back into the recycling bin or was used as a starter for my fire. It's really surprising what a difference we could make just by printing on both sides of the paper. Ashley Steven of NuRelm, an US organisation that runs workshops on how to reduce paper use in offices, estimates that a 12-feet high wall, running from California to New York, could be erected using just one year's waste paper from American offices. When you consider that recycling one ton of waste paper (the amount an attorney in New York gets through in a year) could save seventeen trees, the benefits of reducing our paper consumption even slightly could be huge.

It wasn't quite so easy to find waste pens. There's no obvious place to look, so it comes down to serendipity. Pens (and lighters), are probably the most disrespected products on the planet. When I worked in offices, just a handful of us, forgetting where we'd left the old one and pulling out a fresh one, could regularly get through a box of cheap pens in a month. I benefited from this disrespect, finding pens behind park benches and pens on footpaths, not to mention the half-eaten ones down the back of various friends' sofas. Not exactly a solution all the world could use but whilst things are going to waste, isn't it our first obligation to use them before producing anything else?

Between the solar panels and good old-fashioned handwriting, I met all my writing commitments, though not without the odd expletive. But that wasn't the end of my solar panel problems. Interest in my experiment remained high until

MAKING MUSHROOM PAPER AND INK

Figure 2 Birch polypores – 'you're going to need a stick!'

PAPER

Find some birch polypores (*Piptoporus betulinus*) that are white on the underside and flexible. They can be moist or dry but not dried out. Or you can use old (no longer tender) Dryad's Saddle fungus (*Polyporus squamosus*), even large, maggot-eaten ones.

Get enough to experiment with. The contents of a medium basket will make about 15–20 A4-sized sheets of paper.

Remove the dirty bit where they were attached to the tree and chop the fungi into small pieces. Liquidise with water or natural plant dye (berries, leaves or roots) to the consistency of runny wallpaper paste and pour it into a tray.

Use a paper-making mesh and deckle, or a fine pan cover, to scoop up some pulp evenly across the mesh.

Allow to drip for five minutes.

Flip mesh over on to a fine cloth. Gently press all over with a sponge, to absorb excess water, squeezing it from time to time.

Cover with a towel and press down firmly all over.

Carefully remove the mesh, making sure to hold the cloth down. Allow to dry completely before peeling off the finished paper.

Figure 3 Mushroom paper

INK

Gather some inkcap mushrooms and leave them on a plate for 3–5 days to liquidise.

Strain liquid through a fine cloth and boil to concentrate to half its volume.

Experiment with different colours (using plant and berry juices) and different thicknesses.

the beginning of February; journalists would regularly phone me for comments. This put a huge strain on my phone and its solar charger just wasn't up to the challenge. I often had to charge it via my laptop, which put a further energy drain on it and my battery.

By the middle of February, people had stopped calling me so much. This was interesting; for months I'd received text messages and missed calls that I was unable to reply to. People who knew I was living without money, especially journalists, nonetheless often asked me to phone back, which was frustrating. How they expected me to, I've no idea! My lack of response eventually made me something of a forgotten man. I told myself it wasn't through any conscious decision to ignore me but because I could no longer remind people that I didn't see every week that I was still alive. At least, I hoped that was why!

Between the media interest calming down and the longer days as winter faded away, the problem began to sort itself out. The toughest part of the year was over. I was really looking forward to spending less time putting on and taking off my wellies and more time lying under a tree with a book; to cycling in daylight and to the sense of new life and freshness that we experience when spring arrives!

10

A SPRING
IN MY STEP

In the past, I'd never really noticed the change of the seasons; living in a city disables your ways of reading the signs of such an extraordinary evolution. But living among nature makes you much more aware of her idiosyncrasies. There is definite magic in the change of a season, in the same way the first glimpse of the sun over the horizon signifies the end of night and the break of day. I can pinpoint the exact moment I felt winter was over.

It was the second to last Thursday of February, seven days after the last of the snow had melted. I found myself, for no apparent reason, waking up even happier than usual. About 7.15 am, as I was reading, a ray of the sun burst through a gap in my curtains and I heard a beautiful little song just outside my window. Before long, this had become the most tremendous choir; I felt the birds had spent the winter practising just for me. Later that morning, I walked outside without wellies for the first time since I began my experiment. I even contemplated losing the t-shirt and getting into shorts. And only one week earlier, my caravan had been covered in snow.

Walking around the farm I saw flowers in bloom; snowdrops, rhododendrons and daffodils – for me the embodiment of spring – were showing their faces. But I was concerned to see the cardamine (lady's smock) and euphorbia also coming out; flowers you wouldn't usually expect to see before March. I've noticed them arriving slightly earlier every year since spring 2005. In nature, a few weeks is a long time; this trend to earlier flowering is an indicator of a changing climate.

For the first time since I had begun, that night I cooked my dinner after six o'clock without having to use my wind-up torch. The thought of things getting easier was fantastic. I felt completely renewed by the thought of the longer, warmer days ahead but what excited me most was the start of a new food season. I do love winter vegetables, especially pumpkin, celeriac, purple-sprouting broccoli, turnip, swede, carrots and parsnips. And what Irishman doesn't love a potato? These crops are earthy, heavy and warming on a cold winter's evening. But it was spring, and I could sense the life and energy coming back into my body.

I wanted food that matched my new needs. I didn't want to blast the nutritional value out of my food by cooking it at high temperatures. I craved raw food. Luckily for me, spring is the start of the raw food season in Britain; unless you are happy to eat imported food during the winter, it really isn't very possible then. Now I had wild watercress, wild garlic, cucumbers and salads like lettuce and rocket coming through. Life tasted fantastic again. It's lucky that nature supplies us with this extra energy at the beginning of March, as spring is one of the busiest times of the year when you live off the land. One of the first and most important jobs is to get the wood in.

WIELDING THE AXE

Stocking up on firewood isn't the first job most people think of when they think of what needs doing in spring. You're leaving the cold weather behind and the woodburner gets to take a well-deserved holiday.

But just as you don't have food in autumn if you don't plant seeds in the spring, you won't have a warm home if you don't fell and store wood before the hot summer months. For wood to burn well, it must be seasoned. When you fell a tree, the wood contains a lot of water, as you know if you pick up a fresh log. Leaving it to dry through the spring and the summer means you'll have some really decent firewood in the autumn. If I'd been certain that I would return to my old life in the city once my moneyless year was over, I wouldn't have bothered. I wouldn't need the wood there, as there are strict regulations about burning it in cities. But at the start of the spring, I had no idea whether I was going to continue living without money if I made it to end of November, so I applied the precautionary principle and got it in regardless.

Because the winter months had been so hectic, I'd neglected my wood-chopping activities and didn't really get started until late February. The farm where I volunteered had an overgrown piece of land that hadn't been looked after properly for years. It had plenty of trees overdue for felling and coppicing, meaning ample amounts of wood for me. I went on to the Toolshare scheme on the Freeconomy Community website and borrowed some tools. The people who lent them to me were more than happy to share but I felt slightly uncomfortable. Most people who borrow tools know that if anything happens they can buy new ones. I didn't have that luxury, so I was paranoid about damaging my borrowed tools. However, it did mean I looked after them really well.

The tools I needed depended on what wood I was coppicing and the stage of growth it was at. Coppicing involves cutting down young tree stems to almost ground level, which, as well as giving you some wood immediately, encourages new shoots to sprout. For hazel and other young re-growth, billhooks (a traditional hand tool similar to a machete but with a hook at the end) were the fastest and neatest implement to use. Loppers (scissors on the end of a long pole) and a pruning saw were good for smaller stems and on older coppice I found a bow saw worked best. I managed to get all these from Freeconomy members in Bristol and Bath.

My first job each morning was to get my tools together and select the trees I felt were the best to cut back. This was my favourite part of the day. The rising sun peeped its face over the eastern horizon of the valley a bit earlier each morning, thawing the light frost that carpeted the hills I roamed. The birds engaged in an avian *X-factor*, each trying to outdo the other. The difference was they could all sing fabulously. And the rabbits realised that *Homo sapiens* had woken up and wisely scampered out of my vegetable patch to the safety of the hedges. Little did they know I was vegan.

Cutting down trees has, rightly enough, a terrible reputation. Humanity is chopping them down at an alarming rate at a time when we need more, to absorb the growing amounts of carbon dioxide in the atmosphere. But using your own fuel, grown a few feet from your home, is a much greener source of heating than piping it from Norway or transporting it from war-torn or fragile countries. We don't only need to reduce food miles to avert the worst effects of climate chaos – we also need to start thinking about fuel miles.

Once the wood had hit the ground, usually before lunch, my next job was to chop it up into smaller pieces, to enable it to season more quickly. Different woods (except ash, which you can burn immediately) take different lengths of time to dry out, but a year is enough for most. I didn't have this luxury. With my supplies almost exhausted, I needed to have some ready within six months or I'd have a really cold end to my year. After splitting it with an axe, I took as much as I could into my caravan, where I stacked it next to the woodburner, giving it a chance to dry inside during the last few months of cold weather. The rest I covered with a tarpaulin, waiting for the summer sunshine to dry it. Every day, I took wood from under the tarpaulin to replace the wood I had burned the evening before.

It was incredible how quickly the snow of January faded from my memory. I adored my days out gathering wood, which took me the best part of the last two weeks of February. For the first time in my life, I'd gone topless wood-chopping on St Valentine's Day. Unfortunately for me (but fortunately for them), the only females around were eating grass in a nearby pasture. Claire had presumed I would be too busy to want to do anything. There is something about chopping wood that resonates with something primitive, but still very alive, deep inside. My female friends told me it was a male thing, a deep-rooted need to provide for our partners. Maybe, but after just three months of living without money, things weren't going very well for me on that front.

RELATIONSHIP PROBLEMS

When I tell people that I live without money, the first things that spring to their mind are the physical challenges. However, they are only half the battle. I undertook the year not only to see if I could cope in a 'bushcraft survivor' type of way but also to find how it felt, personally and emotionally, to live without money. And to be honest, it was very challenging, especially at the beginning.

Just before I'd started the year, I'd begun going out with Claire. She was very supportive of what I was doing but not looking to do it herself, partly because she had just begun a degree in Environmental Geography and needed to pay her way through that. She knew before we started going out that I was going to have a really busy year and she was happy to work with that. Practice, however, is always much more difficult than theory. The demands of moneyless living, coupled with the media interest, meant I was constantly busy. If I wasn't doing the things that living without money calls for, I was writing or talking about them. And my decision not to get into motorised vehicles for the rest of the year didn't help.

That decision, in many respects, was ridiculous. Claire often took her dogs for a walk along the coast. This was about forty miles away, well outside my cycling range. But she was going there anyway; the reasonable thing would have been to go with her and have a beautiful day together. However, I felt I needed to make a statement about oil, especially to those closest to me, and it would have been hard for people to take my stance seriously if I'd continued to use oil myself. Understandably, this put a strain on our relationship. She thought I was going overboard and maybe she was right. But I felt I had to stay true to my ideals.

Before we knew it, the little petty arguments, often an indicator of bigger problems, were happening. We cared about each other, and she supported the way of life I was trying to promote, but the realities of going out with someone who has given up most material possessions didn't quite match the romantic illusion, especially for someone who needed to keep one foot in the 'normal' monetary system. The pressures spring put on my time were greater than ever. The weeds suddenly came back to life exactly when I wanted to plant my seeds. So, in the middle of April, Claire and I decided to break up. It was painful for a while, as every split is. Days I should have spent planting the seeds for my summer's harvest, I frittered away feeling sorry for myself and questioning whether I should pack it all in, sacrifice some of my ideals and have long, lazy weekends in bed with a girl I loved. But being moneyless helped me get over it more quickly than normal; I knew that unless I started breaking sweat again I'd not have much fresh produce to pick after June.

This highlighted one of the ironies of my life. I spend most of my time doing stuff for people I never meet, let alone who care for me. Then I neglect those who are dearest to me, because I am too busy with the other stuff. How do you balance your responsibilities to those you deeply, personally, love and care for (who you can usually count on your two hands), whilst simultaneously trying to do your best for the people and the planet negatively affected by the way we live here in the West?

Breaking up in the middle of spring brought other issues. Summer is a time of romance, a time to spend the long, light evenings with a partner. I was back on the market with one of the worst lonely hearts ads you could possibly imagine:

DESPERATELY SEEKING

MARK, 29, BRISTOL.

Single white male, Irish, no money, no car, no television and no career (and little prospects of things changing). Has own house (14ft caravan). WLTM single vegan female, with penchant for moneyless living, into local organic food and permaculture, GSOH and model looks. Can take lucky woman skipping for dinner at weekends, weeding in the evenings and solar showers together in the morning.
Call Mark on 0845 HOPELESS

How I live my life raises some very personal dilemmas. I've chosen this way of life for myself but will a potential partner still be interested if I decide I want to carry on? It can be pretty hard to find someone you adore at the best of times. Vegetarians, vegans and locavores (people who only eat food grown within a defined radius of their home), who decide to go out only with those who eat a similar diet, know how much this decision narrows down the list of potential partners. How much worse might it be if you are looking forward to a life without money? I often jest about it but I would be lying if I didn't acknowledge it weighed on my mind from time to time. Even moneyless people want to fall in love!

And as if things weren't hard enough, my old chat-up line had become obsolete. In the past, if I fancied someone, I'd ask them out for a drink and we'd go for a glass of wine or a cup of tea or coffee. But I hadn't yet brewed any alcohol and as I couldn't hit the local coffeehouse for a double espresso, a cup of freshly-picked wild tea was the only thing on the drinks list for any girl I wanted to impress.

TWO CUPS OF TEA ...

Spring is a great time for foraged tea. My favourite has to be nettle and cleaver tea, as much because they both grow beside my front doorstep as for the taste. They make a fantastic brew, packed full of nutrition and anti-oxidants, high in iron, potassium and magnesium, and with traces of other minerals.

For most people, there are many ways to make a cup of tea: black or white, with or without sugar and of infinite degrees between weak and strong. However, if you consider the whole process of making tea, there are just two. The first is what I'll call the 'sane' way of making tea. I'm assuming the overwhelming majority of the population is sane, so given that this is the way it makes its tea, it stands to reason that this is the 'sane' method, or people would choose otherwise. It goes like this:

1. Get people in India to grow some black tea. Plant it, weed it, harvest it, dry it, then sell it to a local wholesaler for a sum of money on which they find it increasingly difficult to survive (unless it is fairtrade).
2. Send it 4000 miles by air or sea to the UK.
3. Send it to a UK wholesaler or central warehouse by lorry.
4. Transport from the warehouse to a retailer close to where you live, usually by van.
5. Give the shopkeeper about 99p, not a lot when you consider the number of people involved in the process.
6. Bring it home and plug it into the socket, thus ordering the national grid to give you electricity to boil the kettle.
7. Grab yourself a mug and enjoy your cup of tea, perhaps while watching television at home, or maybe outside a café watching the cars go by.
8. Feel awake and alert from the caffeine in your tea.
9. Start feeling tired, in the short term as the effects of the caffeine wear off and in the long term as the tannin in the tea stops your body absorbing certain nutrients.
10. Urinate the tea, its toxins and your nutrients into your drinking water supply via your toilet.

However, there is another way to make tea. This I call the 'insane' method, on the basis that the sane masses choose not to do it this way. It's how I made my tea after the spring.

1. Pick a handful of the abundant tea that grows freely around you. I'm fortunate; my tea grows wild within ten feet of my rocket stove.
2. Pick up some bits of wood lying around to burn in the rocket stove to boil the tea.

3. Light up the rocket stove using this foraged wood and boil the water with nettles and cleavers in it for about ten minutes.
4. Grab a mug and look around you at the stunning landscape whilst you're waiting.
5. Pour the tea into the mug (and some into a flask for later) and enjoy it outside in the country.
6. Feel refreshed and packed full of iron, calcium, magnesium and anti-oxidants.
7. Urinate into the compost heap and activate the fertiliser for your future crops.

It mystifies me that we buy, for example, dried nettle tea bags in a shop for a premium, then, through our taxes, pay our local council to chop down fresh nutritious nettles in the spring! An even better example is the people I'd see passing a huge rosemary bush at the entrance to a gigantic supermarket near where I lived in Bristol, who'd then buy the same herb, dried and in little plastic packets, at high prices! Can we no longer see the food that surrounds us, abundant and free? Or are we so disconnected from nature that we can only see it in a packet on a supermarket shelf?

Not only is wild tea free, it is also much better for you, especially if you make it straight after harvesting and leave it to brew overnight. That way, it is fresh and retains many more of its health-giving benefits. Wild nettle tea helps your digestion if you drink it before meals, is fantastic for your skin, hair and nails and is a perfect tonic if you are feeling physically drained. Given that moneyless living is all about using your body, it is quite useful to keep it healthy!

LESS WEALTH EQUALS LESS HEALTH?

During the winter and until the end of spring, my friends and family were, justifiably, concerned about my health. Not only could I not buy the nutritious food I'd become used to but I also had no money for medicines if I did happen to get ill. My physical well-being was especially important, as I was, for the first time in my life, reliant on my body for my survival. Until the beginning of May, my mum phoned every week, from Ireland, to make sure I was still alive. But I think the fact I had made it to the spring, through the coldest winter of my life, reassured those around me that I might live to tell the tale.

One of the truly great things about the UK is its – free – National Health Service. But this year, I wasn't contributing, and I didn't want to sponge off it. Having said that, I'd paid in for seven years without using it once, so I wasn't exactly dependent on the service. I'm a big believer in being pro-active about health. Putting the best possible food and liquid into your body gives you the best chance of staying as healthy as you can. Because I was going to be even more physically active than usual, I was concerned that I might lose a lot of weight. This probably sounds appealing to those who pay money for gym subscriptions and diet books, to shed the pounds but my battle has always been to keep the weight on. I weighed just under eleven stones (154 pounds) when I started the year and I really didn't want to lose any more.

Contrary to what I imagined, the opposite happened. By early spring, I felt fitter and healthier than I had since my early teens, when I played a lot of sport and I'd gained two stones (twenty-eight pounds). I'd followed a rigorous daily training schedule, from the very beginning, because I knew how physically demanding the year would be and the last thing I wanted was to have to quit the experiment through physical exhaustion. Putting on so much weight by mid-spring was something I'd wanted to do. I believe that no matter what life we live, we are an advertisement for it. People judge the success and health benefits of whatever diet we've chosen both through how we look and how we behave. Unfortunately, society tends to judge on looks these days, so I knew that if I did lose loads of weight whilst living without money, this would send the message that without money, you won't get all the food you need.

This became even more important when I realised I was becoming a public experiment. Living without money doesn't mean you will necessarily either gain or lose weight, any more than does any way of life or diet. In the six years I'd been vegan, people had always questioned me about my diet, usually with a genuine, well-intentioned concern for my well-being. I chose veganism for many reasons, one being that, over time, I'd found it a healthier, more natural diet. But you can be a healthy or unhealthy vegetarian just as much as you can be a healthy or unhealthy omnivore. The same applies to living with or without money.

In years gone by, I'd often picked up a cold around March, when we usually experience a big shift in the weather. This year it completely passed me by, as did the much-vaccinated-against swine flu. An American moneyless comrade, who alternates between living in a cave and housesitting, told me he only got ill when he moved indoors. I have to say my experience tends to agree with his, in complete contrast to what I thought before I started the year.

At the beginning of May, completely through my own carelessness, I did give myself food poisoning. Preparing for a bike trip into the city, I'd grabbed a loaf of bread that had been in the caravan for a few days. When I arrived, I noticed some black stuff on it, which I wiped off, thinking the loaf had rubbed against the soot on the old burnt pot I used on the rocket stove. Big mistake; it was black mould. For the next three days, I suffered. I tried to rest as much as I could but living the way I do means there is always something to be done. The experience gave me my first chance to see the difficulties of doing such an experiment on my own. And it made me wonder what I would have done if my illness had been more serious. Having no money means living without the security that I'd been accustomed to. Even a little money in the bank can buy you time to get back to health but living hand to mouth means you don't have that safety net. Fortunately, I have a lot of good friends and they helped me with my usual tasks. Deasy, the farm co-ordinator, who'd become a really good friend during my first six months, made me a couple of light meals whilst I rattled between the bed and the compost loo. It was an excellent reminder that friends are the best security and that no matter how badly you behave during your life, the good ones are much harder to lose than money.

I have one chronic health problem that I knew would hit me in the last week of spring and make my life hell: I am allergic to grass pollen. Hay fever (allergic rhinitis), which affects millions around the world, is bad enough when you live in a city, where grass mostly grows in the space between the footpath and the road. But this year, I was living in a big grassy field; my move from city to country was like someone allergic to dog hair going to live in a pound. During the last week of spring, all I wanted to do was climb under my duvet and put a wet towel on my head. Not the most productive way to survive without money.

When I was young, nothing had worked against my hay fever. At eighteen, I had a steroid injection, against the advice of my doctor. When the effects wore off, three years later, the hay fever came back worse than ever. Anti-histamine tablets from the pharmacy served only to make me drowsy. In desperation, I looked for alternatives and discovered herbalism, particularly Chinese herbalism. Within one week of taking a Chinese herb, my hay fever had gone. This was my first experience of alternative medicine and I was really surprised by how well it worked. This year, buying Chinese herbs wasn't an option. I had to look for alternatives to the alternative!

Through the organic food co-op that I'd worked for, I met a local bee-keeper who gave me a couple of jars of his honey. Being vegan, I never eat honey unless it is made locally by a bee-keeper I know and trust. Even then, I'll only accept it if they let the bees keep the honey they need and don't replace it with sugar. When bee-keepers do this, for me local honey is no longer local, as it has the food miles of the replacement sugar embodied in it. The honey helped, but only marginally, so I used my blog on the Freeconomy Community website to send out a plea for help. I was inundated with offers of advice and one of them worked. A lady, Grace, advised me to use plantain, a very common perennial weed that grew all around me. Both greater and lesser plantain are rich in anti-inflammatory chemicals. Apparently many people who believe they are allergic to grass pollen are actually allergic to plantain yet, ironically, taking plantain can help reduce the symptoms of the allergy.

And it seems like the problem is only going to get worse. The results of research by scientists at the Center for Health and Global Environment in Harvard Medical School show that higher concentrations of carbon dioxide in our atmosphere also

---

THE GREAT PLANTAIN HAY FEVER REMEDY

After carefully identifying the right plant (use a good wild food book), pick ten to twenty leaves. If you don't get much time for foraging, pick more and dry the rest (you could put them in a pillowcase and leave it on a warm radiator).

Put the leaves in a teapot. Pour some cold water on the leaves first so you don't scald them, then top up with boiling water.

Let it cool and put it in the fridge.

Drink a cupful a day, starting before your hay fever normally starts and keeping going throughout the hay fever season, or until it stops. It can taste a bit earthy, so if you don't like the taste add some squash. I find it fine as it is.

Enjoy your summers again.

---

lead to higher levels of pollen, giving me yet another reason to want to reduce my carbon footprint.

Sometimes, we forget that our minds are part of our body and that the foods we eat affect our moods and level of general happiness. Before I started this year, I'd lived for a year without using oil or any of its derivatives, such as plastic. My diet had consisted totally of organic, locally-grown, vegan food and I hadn't used any oil-based packaging. At the beginning, I'd felt quite down and emotionally low. My body and mind had got used to the protein, nutrients and minerals that China's lentils, Bolivia's nuts and the US's soya supplied and it didn't cope very well when I couldn't find instant replacements. It's not that we can't grow these types of crops here; rather, we've sub-contracted our food security to countries where labour is much cheaper.

In the first month of my oil-free year, my sleep was quite disturbed and I felt weak and unhappy. At the time, I had no idea why but a consultation with a nutritionist revealed that I lacked an essential amino acid, tryptophan. Supplements were out of the question, so I hunted down locally-grown foods high in tryptophan, such as mustard greens, foraged hazelnuts and seaweed, broccoli, kale, sprouted rye grain and spinach. Within weeks, I wasn't just back to normal, I was feeling more energetic than ever and sleeping better. This experience stood me in good stead for my moneyless year and I made sure my diet contained a mix of these foods.

Throughout the winter and early spring, my mental health was really good. But, healthy as it had become, in the middle of the spring it faced its first test.

11

UNWELCOME VISITORS AND DISTANT COMRADES

THE UNWELCOME GUEST

As anyone who lives outside in the colder months understands, mice and rats are never far away, as they attempt to escape from freezing temperatures and be close to a nice steady supply of food. Modern houses are built to make it difficult for rats and mice to get in; it's harder to keep them out of low-impact dwellings.

The question for the outdoor dweller is how to deal with the little housemates you'll undoubtedly attract. Rats and mice will be your chief visitors, though an infestation of cockroaches can play havoc. These three species – especially cockroaches – could probably survive almost anything but of rats and mice, mice are the more pleasant. They're small, they're terrified of humans and their capacity for damage is pretty limited. I wasn't too bothered when one moved into my caravan in mid-February. She had her place in the wardrobe (I never actually ascertained my mouse's sex but the mess it made reminded me of a few of my ex-girlfriends, so I joked that it was female), she kept pretty much to herself, and she didn't give me grief for staying up too late at night.

However, at the beginning of spring, she decided that she wanted to build herself a nest. And she chose to do her DIY at three o'clock in the morning. She'd nibble the insulation out of the wall, drag plastic bags and newspapers down from the top of the wardrobe and generally cause a fair amount of disturbance. In my turn, I'd spend half the next day boarding up the hole she had made, which from her perspective was equally inconsiderate. Without fail, the result was a night of rodent rage; the pocket-sized monster would spend the best part of four hours undoing my handiwork by digging another hole just above it. If we'd continued these tit-for-tat revenge attacks indefinitely, the whole wall would have become a series of little boards. So, after my second attempt to shore up her private entry to my house, I conceded defeat and effectively cut her a set of keys.

If you live with people who party hard at that time of the night you could get up, join them and have some fun. Not the case here, unfortunately. After a few weeks in which I got very little sleep, the critter started to drive me mad. I was happy to co-exist peacefully but this wasn't peaceful. The final straw came when she got stuck into the bag of rye grain I'd spent a day working for. The sack was too big for the press in the caravan and I hadn't then found a large enough metal bin to protect it. Every day, my mouse would make a new hole, and once inside she left her droppings everywhere.

For most people, the solution would have been simple. Get some traps and some poison, lay them down and let time sort the problem. But even if I could have bought either, my vegan beliefs meant I didn't want to. I didn't want the mouse eating my hard-earned rations but neither did I think executing her for theft was a fair punishment for her 'crime'. Those who campaign for greater animal rights say almost everyone acts 'speciesist' sometimes, in the same way some people are sexist or racist. If a shopkeeper finds someone stealing a bottle of wine, and the thief doesn't manage to bolt through the door, the shopkeeper can either phone the police or tell the shoplifter to get the hell out and never return. If it's a one-off, the shopkeeper might leave it at that; if the problem persists, they might invest in CCTV to dissuade other potential thieves. However, if a non-human animal, such as a mouse, steals a few grains of rye, we impose the death penalty, which I personally feel is a bit harsh. They're just trying to survive in a land in which, because humans have manipulated and shaped it, little wild food remains. You could argue that we've stolen their food and they're merely taking it back.

How could I dissuade the mouse from soiling my food? I hid every bit I could and made sure I left no cooked food lying around after dinner. This saved the food but it didn't stop her nightly nest-making waking me up. I tried every possible way to encourage the little thing to move out. A friend suggested I rammed cloth into the holes to plug them up and sprayed the material with peppermint water; mice apparently hate its smell. But nothing worked. I briefly questioned my vegan belief system. And if I hadn't had such a short haircut I'd have torn half of it out by April. After two months, lack of sleep was starting to take its toll. Between living the slow life, writing and talking about it, volunteering, cycling, organising Freeskilling sessions and administering the Freeconomy Community website and its global network of local groups, life was really busy. Waking up at three every morning didn't help.

Just as I felt I couldn't take it any longer, the late spring sunshine came out in full force and the higher temperatures persuaded my unwanted guest to move outdoors. In the end, the answer was simply patience. I no longer had to wash the surfaces the mouse walked and pissed on every day. Peace and tranquility returned.

LOW-IMPACT SHELTER

I got my shelter through Freecycle but I am very aware that this is not a possibility for everyone. It is difficult to talk about shelter as 'free', surrounded as it is by issues of planning permission, land ownership and tax.

However, there are some types of housing which can theoretically be (and often are) free. Even if you cannot find a way to make or locate any of them completely without money, they will cost just a fraction of the price of a 'normal house', enable you to live off-grid for the long term and have very little impact on the environment into the bargain.

The problem with most housing, for me, is that you have to take out a mortgage that you'll spend the best part of your working life paying back. This ties people into the wage economy and, in many cases, into jobs they don't even like.

LOW-IMPACT DWELLINGS

Earthships: I dream daily of living in one of these. The brainchild of Michael Reynolds, an architectural genius, earthships are a type of passive solar home made from recycled and natural local materials. (Passive solar homes are houses designed to use the sun's energy to stay warm in winter and cool in summer without the use of fans or pumps.) Made from old car tyres rammed with earth, beer cans, large glass panes, photovoltaic panels and wind turbines, earthships are self-sufficient in food, water and energy. Fantastic design – glass bottles are used to create stunning lighting effects – makes them visually beautiful, to boot.

Underground houses: Subterranean homes maximise the space in small areas, the excavated materials can be used in the building and they are wind-, fire- and earthquake-resistant. One of the greatest benefits of underground homes is their energy efficiency, as the mass of soil or rock (the 'geothermal mass') surrounding the house stores heat and insulates the house, keeping warm in winter and cool in the summer.

Benders: Wooden frames draped with canvas or another waterproof material. Not exceptionally difficult to design; they can easily be constructed for free from local wood and recycled materials.

Roundhouses: Circular houses, with a frame of wooden posts covered by wattle-and-daub or cordwood panels finished with cob. (These are both ancient building techniques, in which a lattice of woven material is daubed with a sticky mixture, often of clay, soil, dung and straw.) Their conical roofs are usually either thatched or have a reciprocal frame green roof (a simple, self-supporting structure that does away with the need for a central roof support; green wood is used freshly-cut as the water content keeps it flexible and easy to build with).

Straw bale homes: Houses built using straw bales to form the walls of the building. In the UK, the bales can be of wheat, rye or oat straw. They are highly-insulative and can theoretically be made for free and from locally-grown materials.

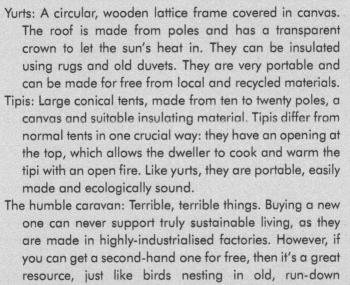

Yurts: A circular, wooden lattice frame covered in canvas. The roof is made from poles and has a transparent crown to let the sun's heat in. They can be insulated using rugs and old duvets. They are very portable and can be made for free from local and recycled materials.

Tipis: Large conical tents, made from ten to twenty poles, a canvas and suitable insulating material. Tipis differ from normal tents in one crucial way: they have an opening at the top, which allows the dweller to cook and warm the tipi with an open fire. Like yurts, they are portable, easily made and ecologically sound.

The humble caravan: Terrible, terrible things. Buying a new one can never support truly sustainable living, as they are made in highly-industrialised factories. However, if you can get a second-hand one for free, then it's a great resource, just like birds nesting in old, run-down buildings. Paint it green to blend it into the natural landscape.

Planning permission is always an issue, no matter which structure you want to go with. Seek advice from your local council. Or, don't bother, just do it and deal with the consequences when they arise!

MY DISTANT MONEYLESS COMRADES

The longer days and increased sunshine of spring gave me more solar energy to play with. Over the winter, I'd known I was receiving lots of emails about my experiment but I couldn't store enough power in my battery to read or respond to most of them. This compounded the slight feelings of isolation I had around the time I broke up with Claire.

As well as charging my battery, I think the sunshine re-energised me. For the first time in my life I was living almost entirely outdoors. By April, I'd developed a tan I normally wouldn't have built up until at least the middle of June. With my body and battery brimming with more energy than ever, I decided to dispel my feelings of loneliness and try to locate fellow moneyless humans out there in the world. As I was pretty sure no one else was doing it locally, I turned to the world-wide web.

Most people who talked to me about my life thought I was the only person in the world living completely without money. I was (as far as I knew) the only person doing it in the UK but by no means the only person living this way in contemporary society. Compared to two others that I became aware of, I was a rookie. Heidemarie Schwermer, the sixty-seven-year-old German author of *Das Sterntalerexperiment – Mein leben ohne geld* ('The Sterntaler Experiment – My life without money'), has being living almost completely without money for thirteen years. (She keeps back a few Euros from her pension each month for train fares; the rest she gives away.) In a film about her experiences, *Living without Money*, she explains how she worked for many years as a teacher and psychotherapist in Dortmund. Like most people, she spent most of her time earning money to buy the things she needed – and also things she didn't really need. As a psychotherapist, she met many people who were depressed and frustrated, over-worked and with very little spare time. Among the unemployed and poor she often found those who thought themselves worthless. I had heard something of Heidemarie before I started my year, through Markus, a friend of mine who spoke German. At the time, her writing was all in her native language, so I wasn't able to explore her thinking and experiences unless Markus translated. However, as the media has become more intrigued by the concept of a moneyless society, she has tried harder to communicate with the English-speaking world.

Heidemarie started an exchange circle, a *Tauschring*, through which people with little or no money could trade objects and favours (*Gib und Nimm*; 'Give and take'). Through the exchange circle, people came in contact with each other in a new way. They felt useful and worthwhile and appreciated the social aspect of their contact. After a while, Heidemarie decided on an experiment. She gave up her flat, donated her possessions to friends and started a new life based on exchanging favours without using money. At the start, she stayed with friends and acquaintances, took care of their houses when they were away on holiday or travelling and in return received food and a place to live. Over the years, she has inspired exchange circles across Germany. The only people she hasn't convinced are the managers of the national train service, which is why she keeps some money for fares. I wondered why she didn't hitchhike but I suppose that's much easier for a thirty-year-old guy than for an elderly woman. Heidemarie's goal was simply to create a 'greater awareness of the relationship to money and consumption'. And whilst she wasn't living completely without money, her exploits were enough to inspire many to use it less and less in their everyday lives.

Another person who has shot into the limelight, as a result of a similar surge of interest in moneyless living, is Daniel Suelo, a forty-eight-year-old American guy from Moab, Utah, who has been living totally without money since 2000. That put my minuscule year into proper perspective. I hadn't heard of Daniel before I started but my research in April made me aware of his blog, which seemed hugely over-looked, given how long he had been living like that. In the middle of my year, when media interest was, thankfully, taking a rest, the well-known American magazine *Details* ran a story on this 'caveman', from which MSN put the story on their homepage. Interest in Daniel went from almost zero (just like his bank balance) to (unlike his bank balance) millions of people overnight. His blog, in a similar manner to mine, became a forum for intense debate. Cynics, without taking a moment to understand his motives, felt it their duty to tell him what a low-life he was. Who would have thought that living in a cave, and having a zero carbon footprint in a world whose climate is changing rapidly, would be such a social crime?

I want to share some of Daniel's thoughts with you, to give you a flavour of his views. They aren't necessarily my views; although I agree with much of what he says, I don't by any means concur with it all. I thought it would be helpful to give you another perspective on why people like Daniel, Heidemarie and I see living without debt, credit and little bits of paper as crucial to our ability to survive and thrive on our planet.

Suelo on 'ownership and possession'

A moneyless existence ... is not a matter of giving up possessions, because there is nothing to give up, really. Nobody owns anything, so it is simply a matter of realizing that you already own nothing. Then, when you lose something and you realize you never owned it in the first place, there is no sense of loss. And when somebody asks you for something, you freely give it to them because it really isn't yours to give anyway. Then have faith that everything comes as you need it in the moment.

Suelo on 'living without money'

To say that I live without money isn't saying anything, really. That's like saying I live without belief in Santa Claus. Now, if we lived in a world where everybody believed in Santa Claus, you might think I'm stepping out on a limb to live without Santa Claus.

Suelo, asked 'Do you think money is evil?'

No. Money is illusion. Illusion is neither good nor evil. Imagine if you had eyes that saw reality rather than your own belief.

> Imagine if you saw a $100 bill as a piece of paper with a pretty work of art on it and nothing else ... One time I found a $20 bill and decided to play with it in this way. I cut it up and made a collage out of it.

The $20-bill incident brought down a lot of negative comment. There was great debate about whether it would have been better to give it to someone who needed it or better to stop assigning it an illusory value and to take it out of circulation. But, by giving it away, would he have been guilty of reinforcing a system that gave rise to the inevitability of some desperate person really needing it?

On the same day that I first heard about Daniel Suelo, I came across a Sioux Indian, John Lame Deer. He summarised how he felt about being made to use money – and hence become 'civilized' – by white men:

> Before our white brothers came to civilize us we had no jails. Therefore we had no criminals. You can't have criminals without a jail. We had no locks or keys and so we had no thieves. If a man was so poor that he had no horse, tipi or blanket, someone gave him these things. We were too uncivilized to set much value on personal belongings. We wanted to have things only in order to give them away. We had no money and therefore a man's worth couldn't be measured by it. We had no written law, no attorney or politicians, therefore we couldn't cheat. We were in a really bad way before the white man came and I don't know how we managed to get along without the basic things which, we are told, are absolutely necessary to make a civilized society.

Daniel Suelo, Heidemaire and I have slightly different core reasons for wanting to live without money. I prefer not to focus on minor differences but rather to see that which we share. A common thread links the three of us: our desires to see friendships grow between local people through the simple act of sharing, and to see the spirits of kindness and giving reign over greed.

There was an irony in my life. Spending so much time speaking and writing about creating friendships through sharing, and the importance of rebuilding the communities we live in, left very little time for my own life! In the middle of May, to remedy this, I decided to start having a lot more fun with my friends. I hoped the imminent return of summer would mean a lot more playtime.

In the first few months of spring, I found I was still counting down the days to the end of my year, viewing it as something to get through, instead of a challenge to embrace. But by May, I found days on end were passing in which I didn't even think about the 'm' word. Only when someone enquired did it enter my consciousness. I loved living off the land but spring was a time when all I seemed to do was work, as everything later in the year depended on how much sweat I spilled then. It was hard to think that the fruits of my labour were still far from ripening. But the time had come to start reaping what I had sowed.

12

SUMMER

Living without money in winter can seem really unappealing but you'd have to be bonkers not to try it in the summer. Long evenings walking in the woods, camping by the beach at the weekend, cooking food that you've grown and picked yourself, cycling, listening to acoustic music by a camp fire, wandering in the wilds foraging berries, apples and nuts, skinny-dipping in the lake and sleeping under the stars. If you fancy trying this way of life for just one season, summer is the perfect time.

The clocks had gone forward. It was officially British Summer Time and I was enjoying the longer evenings. This pleasure is, obviously, not specific to moneyless people. Everyone I know hates to see the clocks go back and I often wonder how we've agreed to something that none of us seem to want. When you cook, wash, work and play outside, cycle everywhere and live off the land, you're ever more delighted to see the sun stay in the sky a bit longer each day. In the winter and spring there'd definitely been times when I felt the effect of not having money: the moment I'd heard my mates were going to see our favourite band play a gig; the time they went to a film I'd really liked to have seen. Now that summer was here, I forgot I was living without money. I simply lived.

Not only did I suddenly have a range of entertainment opportunities – things like camping, which had seemed so much less appealing in the colder months – but life was getting easier in all sorts of ways.

ON MY BIKE

Much as I love cycling, doing more than eighty miles a week in the winter and early spring months was not always fun. Without the right gear, I often got soaked right through. Even if I didn't wear waterproofs, I usually sweated so much that the result was no different. In a change that I loved, because it meant that my lifestyle was becoming normal for them, my friends kept telling me to get a waterproof, breathable jacket. 'With what?', I'd reply; such a thing had proved impossible to find on Freecycle during the wetter seasons.

When it rained heavily, the lights on my bike went on and off randomly. It took me the whole of March to realise this was down to some dodgy wiring in my dynamo, which I immediately fixed. This had made cycling pretty stressful, as I'd suddenly and inexplicably find myself invisible to speeding motorists driving along country lanes barely wide enough for one car. When I heard an engine scream behind me, I'd either have to stand in the ditch until it passed or risk joining the dead badgers and foxes that dotted the side of the road.

But as the summer breezes began drifting through the valley, cycling became not just easier, but something I really wanted to do. Mountain biking is one of my favourite hobbies, so I often went adventuring with friends around Bristol, in places like Leigh Woods. We'd career down the little streams and muddy paths on the steep hills of this huge estate. Off-road cycling is quite similar to life in some respects: if you want to enjoy it, it's essential not to fear falling on your ass.

This pastime was pretty stupid and slightly irresponsible; mountain biking can be a fairly dangerous hobby. Normally, I wouldn't have thought twice about it but my moneyless status meant I couldn't afford to break either my leg or my bike, because I couldn't have paid to get either fixed. But life's too short and too precious to be smart all the time. I decided long ago that I would rather live for fifty years than exist for ninety, and that if I lived my life exactly the way I wanted to at every moment, my time would be called whenever it was meant to be.

However, it wasn't all mountain biking. Cycling is a fantastic way of getting to know the local area and reaching places completely inaccessible to motorists. In June, I started going for cycles in the countryside, sometimes with friends who needed a break from the city and sometimes with the volunteer workers from the farm. There is something very wonderful about feeling the elements and sensing on your skin the change in temperature as the sun falls and rises that makes cycling feel so much more real than travelling by car or public transport. We'd often go cycling at night, which I much prefer to cycling in the day, as you can go for an hour without meeting a single car. As the evenings got longer in the summer, we explored further. We'd shove three things in our packs – tent, sleeping bags and a pannier of food – in case we decided we sleep wherever the road took us. Sometimes we'd camp overnight in woodland, or on the banks of a nearby lake, and make our way back in the morning. Or if we got tired and wanted to lie back and look at the stars, we'd just stop, find a dry spot and sleep until the sun woke us up.

FREE BOOZE!

These days, if I want a tipple I make my own. The world-wide web has hundreds of recipes for all sorts of alcoholic drinks. Or you could try my recipe for cider, easy to make, using windfall apples and nothing else.

HOW TO MAKE REAL CIDER

Pick your apples, a mixture of cider and crab varieties, discarding any rotten ones.

Pulp them or chop them up really small.

Press them, ideally using an efficient apple press, until you have as much juice as you can get.

Pour the juice into a sterilised keg, making sure the keg is full. Release the bung on the top of the keg to let natural yeast spores in.

Allow it to ferment for a month or so. Either pour the cider into clean bottles and leave for another few months or put the bung back in and leave in the keg for eight months. This will give you a strong, sweet and cloudy cider.

Enjoy the cider with friends!

Many people have apple trees and don't use them; why not ask if you can harvest their apples and share the cider with them?

Good beer is also easy to make, especially if you grow your own hops. You can flavour it using almost anything – Andy Hamilton once made me some pine needle beer, which was ... interesting!

THE MONEYLESS SUMMER DIET

One of the many reasons I love summer is the food. Whilst the south-west of England doesn't quite have a Mediterranean climate, in a decent summer you can grow a wide variety of crops and around August, wild food is plentiful. I eat all kinds of different things in the summer. Not all of them every day of course – or I'd have put on even more than two stone over the year! This list doesn't include the stuff I find randomly in bins.

Breakfast

Nettle and cleaver tea	Foraged
or Mint tea	Grown
Porridge oats	Bartered
Blackberries	Foraged
Raspberries	Foraged
Cobnuts	Foraged
Plantain hayfever remedy	Foraged

Brunch

Apples	Grown
Banana smoothie	Skipped
Grapes	Grown
Lemon verbena tea	Grown
or Dandelion root coffee	Foraged

Lunch

Wholegrain rye bread	Bartered grain, ground using hand crank mill, then cob oven baked
or Wholewheat bread	Skipped
Plum jam	Plums foraged and then homemade using grown & pressed apple juice
Margarine	Skipped
Sprouts	Grains & pulses bartered, and then self sprouted

Rocket	Grown and foraged
Lettuce	Grown and eaten raw
Tomatoes	Grown and eaten raw
Oil (preferably from olives)	Skipped
Beetroot leaves	Grown and eaten raw
Grated carrots	Grown and eaten raw
Grated beetroot	Grown and eaten raw
Ramsons	Foraged
Mustard leaves	Grown and eaten raw
Radishes	Grown and eaten raw
Chard	Grown and eaten raw
French beans	Grown and eaten raw
Mange tout	Grown and eaten raw
Onion	Grown and eaten raw
Purple sprouting broccoli	Grown and steamed
Spring onions	Grown and eaten raw
Peppers	Grown and eaten raw
Cucumbers	Grown and eaten raw

Dinner

Potatoes	Grown and boiled on rocket stove
Sweetcorn (on the cob)	Grown and boiled in skin
Courgettes	Grown and steamed
Rye Grain	Bartered, then boiled like rice
Tofu	Skipped and stir-fried
Leeks	Grown and steamed
Lentils	Skipped
Broad beans	Grown and steamed
Leaf curd	Foraged and homemade
Broccoli	Grown and steamed
Garlic	Grown and stir-fried
Carrots	Grown and steamed
Beetroot	Grown and steamed
Pearl barley	Bartered, then boiled like rice
Parsnips	Grown and steamed
Rosemary	Foraged

Parsley	Grown and steamed

Dessert

Vegan chocolate cake	Leftovers from local café

Drinks

Cider	Grown and homemade
Elderflower champagne	Foraged and skipped ingredients
Elderflower cordial	Foraged and skipped ingredients
Apple juice	Grown and juiced
Peppermint tea	Grown
Beer	Foraged and skipped ingredients

Waste food made up less than 5% of my diet in the summer but I didn't stop skipping. I started doing it more and more, partly because I loved the adventure of it and partly because I wanted good food to go into bellies instead of bins.

NO SUCH THING AS A FREE LUNCH?

There is such a thing as a free lunch. And free breakfast and dinner for that matter. Foraging wild food is its truest form, as it comes straight from the earth. However, Great Britain has been tamed; its wilds are retreating rapidly. Where there were once woods, biodiversity and abundance, are now concrete-clad supermarkets, car-parks and their bins. Urban sprawl has changed the nature of 'foraging'. Rather than walking through midday fields picking food, the modern-day urban forager operates at night, searching through the massive bins that have replaced the bushes.

Skipping for food sounds sordid and illegal; I understand this apprehension. But a lot of the time the only reason food has to be thrown out is because of a date stamped on it on an assembly line in a faraway factory. The food may still be fine to eat but the company has to operate within the law. In a small greengrocery, the grocer can judge the state of their produce by its smell, feel, taste and look, and send vegetables for composting only when they're no longer fit to eat. In a large supermarket, layers of packaging mean the assistants cannot use such discretion and judgement. Regardless of how the produce looks and feels within its plastic wrapping, if its date is yesterday's, it's in the bin.

I find bin-raiding a lot of fun, especially if you do it with a few friends. We often come away with so much food that our biggest job is distributing it to those who can use it. Skipping is even easier in the summer, because it is much warmer and drier, two important factors in a night-time activity. And although you have to wait longer for it to get dark, there also seems to be much more waste food around. This is likely to be due to demand being much more unpredictable in the holiday period; sales of many products, like salads and chilled goods, depend on the sun coming out, which doesn't always happen in England.

Bin-raiders are often called 'freegans', although using waste food is only a tiny part of freeganism. A freegan, according to their UK group, is someone who tries to live simply, reducing their consumption and the pressure they place on the environment, through recycling, sharing resources and – importantly – using their time to help others in voluntary work to help positive social actions locally. Some of the most generous people I have met, both of their time and their possessions, are freegans.

But why root around in bins late at night for food that has been deemed unfit for consumption, whether by the law or by a disempowered person along the food chain? For me, I must admit, it's not ideal. The food often comes from industrialised processes, with all their embodied pollution and environmental destruction. And if everyone wanted to do it, there would be nowhere near enough food to go around; the producers would go out of business if no one bought their products. It's hardly a model of future sustainable living, as only a finite number of people could do it. However, not everyone wants to do it. In fact, so few people, that whenever you go to most bins you'll find them full of perfectly edible food. I live near a city of half a million people and I've yet to see a queue at any bin I've been to! I believe that we have an obligation to liberate every pound of edible food from the bins of shops and supermarkets that, for whatever reason, have to throw it out. In 2009, during their food crisis, reports showed Haitian children picking individual grains of corn from the roads, dropped from sacks as lorries drove past. To have good food rotting in our bins is an insult to the families of those kids.

Another reason I feel compelled to use waste food is because once it goes into a bin, it effectively becomes carbon-positive to use it. Not only does using waste food mean that less food, with the embodied energy of its production, packaging, distribution and sale (particularly high for convenience food), is grown and processed, it also, bizarrely, reduces greenhouse gases. Most people believe that because food rots quickly, greenhouse gas production isn't a problem. However, it is exactly the problem. When food breaks down, it produces methane, a potent greenhouse gas. According to Food Aware, every year 18 million tonnes of 'edible' food ends up in landfill in the UK alone (a third of all food, worth £23 billion). That's a lot of climate-changing methane, not to mention the environmental costs of transporting the waste food to the landfill sites and processing it.

From this perspective, you'd think that those who use waste food would be heroes in a world verging on climatic catastrophe and ecological collapse. Unfortunately, the opposite could not be more true. Not only do those who do it risk being socially outcast, it is a criminal offence; technically, theft.

FUN FOR FREE

Emma Goldman, a hugely influential early twentieth-century political philosopher and activist, said, 'if I can't dance, I don't want to be part of your revolution'. Living a life of voluntary simplicity doesn't have to be dull and boring. More often than not it's a whole lot of fun, especially in the summer! Living with money can sometimes seem quite boring: mundanely going for a drink in the pub, a nice meal in a restaurant or to see a film at the cinema. Where's the adventure?

Armed with your home-brew, you're going to want to have a party. One of my favourite organisations is *Streets Alive* (www.streetsalive.net), which advises on how to throw a great street party in urban areas. Not only are they a lot of fun, they're also a fantastic way of getting neighbours out of their houses and into having a good time together, leading to lasting friendships and leaving people feeling refreshed and good about where they live.

In the summer, camping is a superb option. Don't forget to bring your guitar, your drums, something to light the fire with and leave your troubles back wherever you came from.

If you like art, there are always free exhibitions in or around our major cities. Some even have a free bar – I, unfortunately, didn't know this until after I had completed my moneyless year!

One thing there is no end of is free film nights, evenings of films and documentaries for anyone who is interested. If they aren't happening where you live, why not organise one yourself? They're a great way of sharing information and getting like-minded people together. Lots of great documentaries are distributed freely on the web. If you run your evening for free, you should have no problem finding an organisation that will lend you a projector – email your local Freeconomy group.

If music is your thing, 'Open Mike' nights are not just free entertainment but a great way to see new local talent playing acoustic music. If you are even slightly musically gifted, work up the courage and get on stage yourself. The crowds are always supportive and it is a superb way to build up your confidence.

You can find free tickets for all sorts of events on popular websites such as Money Saving Expert and Gumtree and you can get free tickets for many of the BBC's shows. I appeared on Russell Howard's Good News Show on BBC3 and was surprised to find that not only did everyone get to see a top comedy act (Russell, not me) for free, they also got a free beer. I suspect the latter was to ensure that Mr Howard got as many laughs as possible!

Retrieving food from bins on the privately-owned land of the businesses that throw the food out is a legal grey area. If the supermarkets wanted to be difficult, they could charge you with trespassing or even with stealing. In my eyes, if you throw something in a bin, you relinquish ownership. In the UK, no one has ever been charged with stealing stuff from bins, probably because supermarkets realise the negative press coverage they'd get and the can of ethical worms they'd open. Whilst the supermarkets cite legal reasons for the measures they use to stop people like me looking through their dirty laundry, they are a smokescreen; their real reasons are commercial. Every product saved from the bin is one fewer bought in the store. Neither is it in their interests to show us how wasteful their logistics systems are.

Over the summer, I saw supermarkets go to increasingly extraordinary lengths to protect their rubbish. Walls suited to a medieval castle didn't seem to be enough. Some erected wire fences and hired security guards. Some poured blue dye and bleach over the contents of the bin, ripping packaging to make sure the contents definitely weren't edible. If they were truly concerned about the legalities of people getting ill from eating waste food, they certainly wouldn't do this; eating it in this condition would make food poisoning the least of our worries.

Ordinary shopping can feel quite boring. Go in, walk around the shop in the pattern the retail design team have worked out will make you spend the most money, queue for a few minutes, say a polite hello to the assistant, who will probably reply in Head Office-speak, and leave, with your bags full and your wallet empty. In contrast, some of the best times I had during the summer were the nights I went skipping with friends. We'd get on our bikes, with empty panniers (or the bike trailer if we were going to a place where we knew we'd get a haul) and take off for a night's adventuring.

When you go skipping, you've absolutely no idea what you'll find, you just know you'll find something and often lots of it! I've had some hilarious times in bins. The funniest was when I found a case of condoms whose packaging had been water-damaged but whose insides were absolutely untouched! That solved one hell of a headache; Fergus's idea of making them out of the intestines of roadkill badgers was a bit hard to stomach. Not sure it would have aroused future lovers, either. Coming a close second in the 'most bizarre skip finds contest' was my mate Dave Hamilton, who found a £10 note one night and used it in an organic food store the next day to pick up some fine-quality food! Not exactly Freeconomic living but he didn't complain. I found cases of beer past their 'best-before' date (fit for consumption but not in its prime) and cases of wine in which one bottle had broken, meaning the rest were stained and not fit for sale.

Most nights we'd find anywhere between ten and twenty loaves of bread, and stumbling on cases of fruit and vegetables was far too common. But what to do with this excess? We did what most other freegans do; distributed it to friends and others whom we knew would really appreciate it. Many of my friends spend much of their time on voluntary projects, meaning their incomes are minuscule relative to the amount of work they do each week, so they always appreciate any food help. On my journeys back to the farm I'd drop off little parcels, the contents varying depending on whether the recipients were vegan, vegetarian or omnivore. Some even fed their dogs for free from my parcels; the dogs much preferred it to the stuff they usually got from a can.

One of the strangest evenings of skipping came making a film for *Guardian online* with their journalist Jon Henley and cameraman Mustafa Khalili. I'd been reading Jon's fascinating articles for years; jumping into bins with him seemed surreal. I'm sure it was odd for him too. He told me that the following evening he would be dining with the French ambassador in London, as his wife was a journalist for a French newspaper. Yet here he was helping me take pizzas and pies out of a bin. To my complete respect, he really joined in and even took some food back to London himself, including fruit juice for his child. If someone had told me ten years ago that I'd be living without money and rummaging through skips with one of my favourite journalists, I wouldn't have known whether to pack in my degree or study even harder!

My main purpose for skipping during the summer months was to distribute food to others but there were times when I found it particularly useful, most notably in the days leading up to festivals. I had no end of fresh vegetables available at the end of June but a lot of it was stuff that would quickly go off in a hot tent, not great for five-day festivals. I needed processed food that wouldn't go mouldy and required very little cooking, so I'd spend two or three nights before a festival looking for tinned beans, bread, spreads, fresh and dried fruit and snacks to enhance my staple diet of porridge oats, nuts and berries.

THE FESTIVAL SEASON

The south-west of England, where I live, is a festival Mecca in the summer. Best-known for the Glastonbury Festival, there's a different festival every week from May to October in this part of the world. As you can imagine, it's quite tempting to go to as many as you can, especially when they're on your own doorstep.

At the start of my moneyless year, I assumed I wouldn't be able to go to festivals in 2009. First, I couldn't pay to get in. Second, once in, everything costs a lot. Food is the only essential item but I like to have a couple of drinks when I'm listening to music with friends. Third, though some festivals take place relatively close to me, most involve at least a 120-mile return journey, not much in a car but quite a lot on a bike. The distances meant if I wanted to go to a festival, not only would I have to take four days out of my really hectic schedule, it would take two, quite physically demanding, days to get there and back.

The first two problems sorted themselves out in May when Paul Crossland and Edmund Johnson asked me to help them promote their new project, Freelender, at the Buddhafield Festival. In return for helping them, I'd get into the festival for free. I am pretty fussy about what projects I'll support but Freelender (www.freelender.org) fitted the bill perfectly. Its aim was to maximise use of resources in local communities, through a website where people could borrow and lend stuff (from books to bicycles) to and from those who may not otherwise be able to afford them. Not only does it save people money, it makes better use of limited resources and helps build resilient communities through acts of kindness and trust. Very similar ideals to the Freeconomy Community and a good example of an organisation springing up to fill another part of the 'gift economy'; a social movement in which goods and services are regularly given without an explicit exchange agreement, relying on informal custom and the culture and spirit of giving.

I wasn't sure if Buddhafield would be my type of festival. Much as I wanted to help get Freelender off the ground, I was concerned there might be too much Chai Tea and Tai Chi for my taste. But Paul and Edmund offered to get me in and make sure I didn't run out of food for the five days, so I decided to go. During the day I worked in a tent handing out leaflets and questioning people about their attitudes towards borrowing and lending. We ran a Freeshop from which people could take things they needed for free and leave things they didn't want any more, set up a borrowing and lending service for things like blankets, wellies and so on and organised liftsharing so that people could get home from the festival for free.

I had lots of fun in the evenings, despite the absence of cash. I hung out with friends I hadn't seen properly in years, spent time with people I'd met in Bristol but had been too busy to get to know, went for saunas and listened to amazing musicians, favourite bands like Seize the Day. This did me a world of good. Until the end of June, I'd worked seven days a week and although I was working at the festival, the change was definitely as good as a break. But while I greatly appreciated and needed the fun and relaxation, the festival conspired to have a much greater purpose in my life.

It started just before four o'clock on the next-to-last day, when I bumped into a friend, who told me about a great workshop being run that afternoon by a poet, Paradox, who'd given a fantastically inspiring performance the night before. In the workshop, we asked ourselves that common question 'what is the meaning of life?' and how this meaning changed throughout life, with the intention of writing a tragicomedy of our existence so far. It is easy to write off such exercises as 'hippyish' but I feel that we in the modern world spend too little time contemplating our place in the world and where we are going. We began by agreeing that life has no one meaning in itself but that through time, 'I' or 'you' attach changing meanings to our lives to give us focus and a reason for being. I realised that between five and twelve, being the best-behaved boy in school gave me meaning, from twelve to sixteen it was being good at sport and at sixteen to twenty-one it was beer, girls, designer clothes and money. But between twenty-one and twenty-six I obtained my meaning through unconditioning my mind and deconstructing the lies I'd been fed about the world. Now, my meaning is derived from striving to use everything I've learned to be as gentle and respectful as I can towards the planet and all that dwells on it, to make up for the fairly consumerist way I lived the first twenty years of my life.

Paradox read us an extraordinary, true poem. He'd had the most insane life, including quite a few times when he'd been homeless. One day, two women, unknown to each other, had told him they were pregnant by him. Another day, an accident in Mexico, that tore his leg off, left him on his deathbed, due to a lack of blood of his group. One of the reasons he called himself Paradox was because this event was both the best and worst moment of his life. He did lose his leg but he would have lost his life if a group of Mexicans, strangers to him, hadn't tirelessly sought a blood donor by putting out an appeal on all the local radio stations. Lying on his bed, he realised his life until then had been about himself and his ego and he was not the person he wanted to be any more. He decided he wanted to inspire others and to be of service to the world by sharing what he did best: his poetry. One of the last things he said was that, on your deathbed, you'll recognise the things that are truly meaningful in your life and what is important. And it won't be what brand of trainers or shirt you wear or how much you earned last year. It will be your family, your friends and even nature herself.

At the end of the workshop we read the poems we had written. I felt nothing but complete admiration and empathy for the people around me, many of whom I may previously have thought were strange, weird or idiotic. When we heard each other's life stories we realised just how amazing it was that we were even alive! I left feeling completely inspired and I looked back at the last few months with some doubts about the person I was becoming. Finding it hard to watch the destruction and suffering that we humans cause, my way of dealing with it was to become judgemental when I had absolutely no right to be. If I could blame someone else for it all, I didn't have to change my own ways.

Inspired by Paradox, I resolved I would live every day like it was my last. However, I had no idea that the universe was going to make sure I got the message loud and clear the next morning. I had cycled the fifty-five miles to the festival; a six-hour journey during which I'd seen no other cyclists but lots of dead badgers, rabbits, foxes and birds. I was about two miles into the journey home when I heard a car scream over the peak of a small hill behind me and, seconds later, its horn frantically blaring. I scowled over my shoulder: 'I'm in as far as I can go!', only to see the car in mid-air coming straight for me. The momentum of the bike kept me going and the car landed dead, bent in two in the ditch, with a sound like a bomb going off. It came to rest about six feet from my back wheel, so close I could see it out of the corner of my left eye. If I had cycled one second later or if the car had bounced back out of the ditch, I wouldn't have been living without money. I'd still have been without money, just not living.

I made sure the driver was OK. Miraculously, she stumbled out of her car, albeit in complete shock. I had to hold back the tears as I raced up the road on my journey home, so shocked and pumped full of adrenaline that I got home in just four hours. Going through my head was what Paradox had spoken about the day before: how would I live if today were my last? Would I be happy with the last thing I had said to someone? What had I spent my last hours/days/weeks doing? Had I told the people I treasure how I felt about them? Had I wrongly judged a person about whose story I had no idea? Was I the person I aspired to be? The answer, in June, was 'no'. I was living without money and my actions were closely aligned to my beliefs. But for me, that is only one part of the entire solution.

Most people claim to want 'peace', without really knowing what that means. Peace isn't going to fall down on us from above; it is a mosaic whose pieces are our daily interactions with each other and the planet. My personal interactions were, all too often, far removed from the true meaning of peace. I moaned about being too busy, complained about people buying stuff I didn't agree with and generally acted less positively than I could wish. Moneyless living had begun as a means to a more peaceful way of living but had become an end in itself, just like money started as a means to easier transactions but became an end in itself. Paradox's workshop helped me check myself and get back to my original intentions.

The second festival I managed to get to was Sunrise Off-grid, the little sister of a much bigger festival, Sunrise Celebration. This was the off-grid festival's first year. It grew out of the founder of Sunrise, Dan Hurring's, desire to take issues such as climate change and peak oil seriously and to show other festival organisers how to put on a really fantastic festival, have a lot of fun, yet cause very little impact to the environment. Dan had got in touch in May, to ask if I would do a couple of talks on living without money and I gladly agreed. A couple of two-hour talks was much easier than the five days' work I did at Buddhafield, although I helped in the alternative economics section in my spare time.

Sunrise Off-grid was four days of workshops on every aspect of society, from economy to ecology, education to energy, food to friendship and politics to pottery. In the evenings, it was all about music and dance.

That is a point in itself. We've never had more money or cheaper energy. If environmental destruction made us happier, that would be something. Frying the planet would have some justification if it made us joyful. But why haven't we become happier as we've become financially wealthier? Richard Easterlin, an economist at the University of Southern California, believes that a large part of the problem is the consumerist treadmill we are on; never satisfied and always wanting more. He says:

> People are wedded to the idea that more money will bring them more happiness. When they think of the effects of more money, they are failing to factor in the fact that when they get more money they are going to want even more money. When they get more money, they are going to want a bigger house. They never have enough money but what they do is sacrifice their family life and health to get more money.

The Austrian millionaire businessman, Karl Rabeder, realised this simple fact and gave away everything he owned, including his £3 million fortune. Asked why, he said:

> Money is counter-productive – it prevents happiness to come. For a long time I believed that more wealth and luxury automatically meant more happiness. I come from a very poor family where the rules were to work more to achieve more material things and I applied this for many years. But more and more I heard the words: 'Stop what you are doing now – all this luxury and consumerism – and start your real life. I had the feeling I was working as a slave for things that I did not wish for or need'.

I had the privilege of meeting a few of the people who've really inspired and influenced me. One, Patrick Whitefield, Permaculture guru and author of *The Earth Care Manual*, came to one of my talks. Knowing someone in the crowd knows significantly more than you about almost everything you are talking about can be a little unnerving, to say the least, but thankfully he was entirely supportive.

I also went to an interesting talk by the founder of the 'Transition' movement, Rob Hopkins. Rob's talks are always intriguing but this one was particularly fascinating. He had to limit his presentation (for which he used a data projector) to just an hour. If he hadn't, the band due to play a little later on the same stage wouldn't have had enough power for amplification. This highlighted the implications of taking responsibility for your own energy needs. When his talk ended and questions began, Rob turned off his wind-powered laptop, whereas ordinarily, plugged into mains electricity, he'd have left it on. Whenever you produce your own of anything, you don't waste a drop.

I went to a workshop led by Theo Simon, lead singer and lyricist of Seize the Day, one of the bands I'd been delighted to see at Buddhafield. Theo has spent twenty years writing and performing songs that have inspired activists in the UK and beyond to keep campaigning for social justice. Constantly on the frontline himself, Theo had spent much of the summer of 2009 campaigning with the workers of the Vestas Wind Turbine factory, on the Isle of Wight, who had lost their jobs because their bosses, who'd made a profit of £76 million in the previous three months, realised they could make more if they moved their operations to the US. Workers who'd joined just months earlier, advised that their jobs were safe, had taken out mortgages, only to be given barely any notice and just £200 in redundancy payment. Others had lost jobs they'd spent their working lives committed to. That's green capitalism for you: a symptom of a system based on competition instead of co-operation.

Theo's workshop was called *Conscious Activism*. Over the years, he has seen a lot of brutality, mostly from police ordered to defend the interests of those who bring millions of pounds into the UK economy. As anyone who has been on a protest, direct action or non-violent demonstration to stop an incredible injustice knows, some police officers can be very heavy-handed. During the workshop, Theo described many of the incidents he's witnessed and it wasn't easy listening. Every single person in that workshop was touched to their core by how he talked about the police, despite his harrowing experiences of them. Activists often talk like they 'want to save the earth'. The earth will be fine, in time; it's humanity that may need saving. But who do they want to 'save' it for? Only other activists? Only for activists and the working classes? Or for everyone: executive bankers, environmentalists, police officers, human rights activists and politicians alike?

ACCOMMODATION FOR FREE

The way we live today means we often have to travel. But you don't need to pay for accommodation when you get there!

In the country, there is always the tried and trusty tent but in the city this usually isn't an option (though I did wake up on urban football pitches a few times during my year!). And depending where you live, camping may be just a summer option.

A number of great websites look after this department of the moneyless movement. I've found the best to be

Couchsurfing (www.couchsurfing.com), which matches couches available to people who need them, in almost every town on the planet. Not only does this mean accommodation for free, you get to make new friends and access local knowledge of where to go in whatever part of the world you find yourself. I met one of my closest friends, Sarah, when she came to stay on the couch of my old houseboat for a few weeks.

I love couchsurfing, because it is based on a 'pay-it-forward' ideology. It has proved hugely successful but, like Freeconomy and Liftshare, it depends on you helping a stranger for free when your turn comes round.

Others sites include Hospitality Club (www. hospitalityclub.org) and Global Freeloaders (www. globalfreeloaders.com), which both work in a very similar way to couchsurfing.

If we are genuinely interested in preventing the worst repercussions of climate change and depletion of resources, we need to engage with and have compassion for everyone, not just those who have similar views to our own. Turning things around environmentally will have to involve everyone, including the police officers ordered, by their bosses, to prevent such change happening but who, for the most part, do a fantastic job of cleaning up the mess that society creates.

This was particularly relevant to me. I wanted people from all walks of life to get involved in the Freeconomy Community, not just the usual suspects. Again, I realised that my thoughts were no more right than anyone else's; merely another opinion to throw into the melting pot of life.

The festivals were a time of fun, friendship and change, but they also reminded me of a number of important lessons I'd forgotten along the way. And they gave me a great chance to promote the Freeconomy Community. I was amazed at how many people approached me, after the talks I gave, with stories about how they've used it and the friends they met through it.

USING THE LOCAL FREECONOMY COMMUNITY

The Freeconomy website took quite a lot of my time that summer; writing about my year had given it a lot of exposure. But it wasn't all one-way traffic. I used it myself both for giving and receiving: I shared camping equipment with a girl who was going cycling for four weeks in August and built an integrated cash flow forecast and profit and loss account spreadsheet for a member who worked for a local charity bidding for funding to continue its work. This case was particularly important for me. Ironically, I was helping the charity acquire financing but I'm a realist as well as an idealist; I knew the charity couldn't survive without money at this stage and, without it, wouldn't be able to continue its great work for the children of Bristol and Bath.

I also received a number of times through Freeconomy. I learned how to use a cut-throat razor, which by summer was an essential skill: the beard that had kept me warm through the winter had long since outlived its usefulness. And notable help came when my laptop broke down. Unless I'd found someone willing to give me an old one, I wouldn't have been able to continue to raise awareness of the issues that laid the philosophical foundations of what I was doing or administer the Freeconomy website. But it just so happened that the following week's Freeskilling evening was on 'How to make a computer'. Ben Smith, the Freeskilling teacher for the evening, offered to put one together for me. He also showed me how to install Linux as its operating system. He didn't just offer this help to me, he offered it to everyone in the class, together with continuing, free, support for those who needed it. Ben wasn't really anti-Microsoft, just very enthusiastic about people using free and open source software. Thanks to Ben, I got back part of my ability to communicate with the world.

Without the support of the Freeconomy Community, it would have been much more difficult to complete a year without money. But that's exactly the point; it shouldn't have to be something you do alone or a life that must be difficult. As new projects like Freeconomy, Couchsurfing, Freegle, Freecycle and Liftshare emerge every year, living without money is getting easier and easier. And if I can do it, anyone can. I am, genuinely, one of the least talented people you are ever likely to meet.

I had a fantastic time during the summer. Though I was usually on the go from five in the morning to midnight, almost every day, it really didn't feel like work and play were two different things. I loved what I did during the day and had an even greater appreciation of the time I managed to share with friends at night. Many evenings, a bunch of my musically inclined friends gathered around the campfire. Alex played the fiddle, Wally strummed the guitar and we all sang and danced until the temperature told us it was time to cover the embers and get into bed. It did make me think how much easier my experiment would have been in a country like Spain, with more sunshine all year round. But running off to continental Europe would have slightly missed the point; models of sustainable living are needed in the UK just as much as anywhere else.

The summer solstice passed. Whilst many people celebrate this day, I absolutely hate it. After the solstice, the days get shorter and shorter, and I had come to love the warm evenings that seemed to go on forever. But, though my first moneyless summer had come to an end, some of the best days were yet to be.

13

THE CALM BEFORE THE STORM

My tiny collection of eight CDs, nestled in a nook above my bed, had gathered a thick layer of dust. I could never quite work out why I continued to hold on to them. Most were albums that composed the soundtrack to my teenage years; I guess I kept them to cling to a time when all that mattered was that gorgeous girl down the road and who United were playing at the weekend. I also suspected I was holding on to them against the time I might re-enter the world of sockets, those unlikely gateways to the beautiful realm of Ziggy Stardust and the Spiders from Mars.

It had been nine months since I'd played those discs. Nine months since I'd bought my mates a drink or taken a train ride to the coast. But there was a strange sense of comfort in the dust. It was a sign of time passing. Every tiny layer added meant another week closer to accomplishing what I'd set out to achieve.

It was autumn. The longer it went on the less I cared about making it to the end. The finishing line wasn't so appealing. In fact, the thing that weighed most heavily on my mind was the thought of going back. I had let go of so much mental and physical baggage and never felt so liberated or so free. What would I do? Would I go back to a job in the city, get a nice new flat and slowly drift back into a 'normal' life? Or were the slopes I had climbed for nine months merely the foothills of an entire mountain range?

After a coolish summer, the sunshine eventually came towards the end of August, which pleasantly coincided with a brief lull in the chaos of my life. Although I appreciated the gift of life a bit more every day, I was tired. Whilst my friends took their summer holidays abroad, I kept the weeds down. For rest and quiet time, a couple of days away in the woods was as good as I'd got. I decided to make the most of the fantastic autumn weather that England enjoys and take some time out. The end of my year was rapidly approaching and I had a feeling that the slow life I was advocating to journalists would suddenly become the fast life again. I had to make big decisions about what I was going to do after my year was up; I needed thinking time, which thus far had proved elusive.

Autumn is, without doubt, my most cherished time of year. The sunsets in September are wonderful; on clear evenings, my entire valley looked incredibly rusty. The birds too seemed to realise it was their last chance to have some fun; the swallows that lived around my caravan spent the last few hours of light immersed in a ritual dance only they understood. One evening, out for a short pre-dinner walk, I had to stop in my tracks, as hundreds of these little creatures flew chaotically around me, sometimes just inches from my body. The swallows' dance seemed to go on for hours. At moments like this I really appreciated how privileged I was to live this way, in stark contrast to a commute through inner-city Bristol on such an evening.

And autumn is the perfect time to go adventuring. My love of both camping and foraging meant that, for the next month, there would be many more days when work and play remained one indivisible whole.

WILD FOOD FORAGING ADVENTURES

I'd decided to pack in as much camping and foraging as I possibly could. In September, every possible gap in my diary was filled, heading off with one mate or another on a long walk or ride into the English wilds, armed with baskets and bags for gathering food. This also proved to be a refreshingly successful alternative to going to bars or restaurants on dates. I was surprised by my rate of success in asking women out to take a break from the norm. This was strangely life-affirming, reassuring me that not everyone was interested only in how much I owned or earned. It gave me hope that somewhere out there, amid the fields of mass consumption, stood the Moneyless Woman, searching the horizon for her Pauper Charming. I wasn't convinced many of them would be up for it anything other than very part-time, but the hope sustained my weeding.

One of my September foraging trips was a completely last-minute decision to go camping with fifteen friends for a long weekend of food, fun, fire and friendship. We grabbed an Ordinance Survey map and spun a bottle to see what direction we'd take. It pointed us west. This particular short break wasn't about the destination, in the way that previous holidays abroad had been. The journey itself was the holiday; where we decided to rest our heads for the night was almost irrelevant. The journey's beauty lay in its effortlessness. Because it was so spontaneous, we had little time to get any food together, though I harvested as much as I could carry to share with the others. But the experience wasn't really about getting the food together *before* we went; it was about gathering it *as* we went. We picked food as it appeared along the hedgerows and fields that enclosed the paths we wandered.

Many of the group really wanted to find out about the varieties of edible fungi. When you speak to people about foraging, the first thing that usually comes into their heads is mushrooms. Mushrooms, in some respects, have a terribly unjustified reputation; the vast majority of fungi are safe to eat. Having said that, pick yourself a handful of Ivory Funnels instead of Scotch Bonnets (both often grow in the same place) and you'll have an uphill battle to survive. Just a forkful of Death Cap mushrooms (a fungus I threaten Fergus with weekly if he doesn't teach me a new skill) can kill an adult. This sounds a bit scary but it shouldn't be. I have very little idea what I'm doing and I am still alive. So it made a lot of the crew very happy when we stumbled on a giant puffball amongst a field of nettles. For many it was their first puffball and because of its size – as big as a football – everyone was really excited. It was big enough to feed all of us for lunch; absolutely delicious fried with olive oil and garlic.

Another much-loved fungus we found along the way that weekend was chanterelles, a yellow mushroom that smells ever so slightly of apricots. Our experience of finding these was much the same as Dorothy Hartley's, in her book, *Food of England*: 'You find them suddenly in the autumn woods, sometimes clustered so close that they look like a torn golden shawl dropped amongst the dead leaves and sticks'. Chanterelles can be hard to see, camouflaged in the leafy carpet of the forest, but their taste makes it well worth keeping a careful eye out. We also found field mushrooms and blewits (a common fungus in grassy pastures), adding even more flavour and texture to our evening's meals. But we didn't plan to live only on mushrooms for four days. They wouldn't have sustained us for the twenty-five miles we walked each day. We also needed to look for anything with a high protein content: the most obvious source was nuts.

The most abundant, and most edible, nuts on our route were hazelnuts. Hazelnuts are really expensive in the shops but free, in large quantities, if you know where to look. Not only that, they store well; if you start looking in September, and beat the squirrels to them (though don't forget to leave them some), you can easily keep yourself in good-quality protein for a year. I grabbed an extra load, to start my winter stock, but they didn't last long as my so-called friends got stuck into them as we made our way through the woods. What kind of people steal food from the mouth of a man without a penny to his name? We came across some walnuts, a bit too young and wet to taste very good. We found acorns everywhere but these weren't much use; their high tannic acid content makes them taste incredibly bitter. However, if I'd been prepared to bring them home and put in a bit of processing, I could have made delicious acorn bread.

Foraging expeditions can take a lot out of you. Maintaining our energy was absolutely essential, especially if we wanted to have fun in the evenings after we had pitched our tents. I'd brought tubs to collect whatever berries we could find along the way: gooseberries, raspberries and blackberries were the most abundant. We'd fill our tubs, eat the berries, and refill. This truly was 'food on the go'. In supermarkets, conventionally farmed raspberries can cost as much as £1.50 for a small punnet; organically grown raspberries even more. On our walk, we'd sometimes pick a tub of glorious wild raspberries in just a few minutes. It felt like we were getting paid to travel. The cycle path that makes up a quarter of my journey to Bristol and back was full of berries between July and September. If I saw a cluster of really juicy ones, I just couldn't help myself. My purple-stained fingers gave me away whenever I showed up late for meetings, trying to use the terrible traffic as an excuse!

We'd set up camp every evening no later than six, get the fire going and cook the fruits of our labour under a full moon to the sounds of acoustic guitar, violin and African drums. We danced, we sang and we eventually slept, some beside the fire. If there'd been anyone living nearby, I daresay they'd have complained. But that was the point: there was no one nearby. This was the closest to real liberation I had experienced all year. I feel eating food as you wander pushes some deep ancestral buttons and, although I struggle to work out why, I'm at my most alive out in the wilds, picking the food that nature freely supplies, before falling asleep under the stars.

The actual food is but one aspect of the wild food foraging experience. It's also a great excuse to spend time with friends away from the stress of modern life and the relentless sound of cars. Foraging combines everything I adore: being immersed in nature, adventuring, exercise, great food and – if you manage to convince your friends to come along and camp out – a bit of a party to boot.

Weekends like this were a great antidote to my 'normal' life which, if it weren't for my passion to spread the philosophy behind moneyless living, would have been like this all year round. I also found it invaluable to help me stay grounded when all else around me could have started to send me a bit mad. If it hadn't been for this moneyless break with my friends, I think I would have probably done just that.

A WEEK OF COMPLETE SILENCE

Two weeks after I got back from my last foraging trip, I decided I'd have a week of silence. I didn't think this would be much of a challenge, though I did find it hilarious that for one week I'd give up both speech and money. If someone had told me that ten years ago, I would have choked on my big greasy burger.

I wanted to regain control of my tongue and become more aware of how I express myself through my actions. My year had been intense in so many ways, from media and public interest in the experiment to the daily realities of my existence; my life had

FASHION FOR FREE

If we resolved to stop producing clothes right at this very moment and learned to share and mend, my guess is we'd have enough clothes on the planet to last us for about ten years. This decision would give the soil a much-deserved rest. For example, 25% of all pesticides are sprayed on cotton; a massive monocrop that covers the lands of many nations.

I've found the solution for clothes can be the same as for books; necessity being the mother of invention. You have clothes others like and they have clothes that you like: why not organise a clothes-swapping evening and get people together to swap their wardrobes? Everyone comes away with something 'new', not a penny is spent and not a drop of energy used.

If you don't feel confident to do this yourself, then help is at hand – two online organisations, Swishing (www.swishing.org) and Swaparama Razzmatazz (type it in your favourite search engine but I recommend Scroogle!) are on hand for guidance.

Second-hand and charity shops are great for clothes; a brilliant way of recycling and supporting what is often a very good cause. However, although they are very cheap they're not free. I recommend holding a Free Shop (perhaps once a month to begin with), at which people can give and take whatever they want with no need for money to change hands. Get in touch with your local Freecycle group to see if any of their members already organise one.

And don't forget, a stitch in time saves nine. Learn how to repair the clothes you love before they go so far they can no longer be salvaged.

changed a lot. I'd started to dislike the person I was becoming; amongst other things, a person far too loose in his speech. I'd criticised people for things, when I'd done much worse in the past. I'd heard myself saying things tainted with the intention of making me look good, impressive, a person others would want to be around and attracted to. I thought I'd best shut up for a while and take a good hard look at myself.

During a foraging trip in August, I'd realised that I sometimes use speech to replace action and other, more genuine, forms of communication. I'd never hesitated to say 'I love you' to a partner, and whilst I'd often genuinely meant it, I'd also said it out of laziness or a desire to manipulate the person I claimed to love into giving me what I wanted. If you take away speech, you've got to show the person you love them. Much more difficult, though a lot more sincere. Telling someone you love them is a fantastic compliment but a terrible primary source of reassurance; all too often, the words lack depth and substance.

On the heath, Lear asks Gloucester, 'How do you see the world?' Gloucester, who is blind, answers, 'I see it feelingly'. I see it feelingly. I wanted to start 'seeing it feelingly' more. Ours has become a very intellectualised culture, in which those who display a strong intellect are admired, whilst those who feel and understand things instinctively get much less recognition. I found myself falling into the former category. In interviews and articles, I could only justify why I was doing what I was doing intellectually. The fact I simply felt using money went against all my instincts was not something I felt I could use to argue my case. Yet, in my experience, 'feeling' is often much closer to truth than 'knowing'. I could give you a presentation of why organic methods of farming are more ecologically sound than conventional agriculture or I could take you to an organic and a conventional farm, say nothing and leave you to let your heart decide which one makes more sense.

It was hard for the first few days. The longest I'd ever managed to stay silent was probably the same as my longest night's sleep. Not responding when people spoke to me was mentally draining; my natural urge is to give my opinion on anything and everything. And I was no budding Buddha. I had rarely meditated in the past and the few times I had I'd been thinking about all the things I should have been cracking on with rather than focusing on my breathing. I do think meditation is beneficial; a very useful tool in a more conscious and aware life. It is just that I've never been very good at it!

It was interesting to observe how people interacted with me. On Monday, people spoke to me a lot and engaged with me regularly. The same on Tuesday. But by Wednesday, I felt people were talking much less to me, probably because people prefer to talk to people who respond and laugh with them. This made me think about how it must feel to be deaf or mute in a world in which everyone else can hear and talk. Or how it feels to live alone in a city, yet surrounded by people. I felt lonely at times and finished my silence with a greater sense of empathy towards those whom society doesn't seem to value.

The week did wonders for my self-discipline; a tool I find needs constant sharpening. The beautiful thing about self-discipline is that when you practise it in one area of your life it can be easily transferred to another. Siddharta, the hero of Hermann Hesse's classic book of the same name, when a potential employer asks him what his skills are, says: 'I can fast'. 'Fasting?' says the employer, 'what good is that?' Siddharta replies: 'If a man has nothing to eat, fasting is the most important thing he can do'. Giving up something that you are free to do builds character.

What did I learn from my silent week? That it's definitely much harder, if not impossible, to criticise people when you can't speak. That being unable to jump in with knee-jerk reactions to something I didn't like saved me hurting people's feelings. And that whilst I found a week without speaking to be very beneficial and something I'd recommend, I had absolutely no intention of continuing with it.

However, when it came to making a decision about whether or not to continue living without money past the end of my official year, I wasn't anywhere near as decisive. And I had only a few weeks to go.

MEDIA STORM 2.0

Things were about to go crazy again in moneyless world. With only a month of my experiment left, I had been expecting another surge of media interest. The day after my week-long vow of silence ended, I got an email from Adam Vaughan, one of the editors of *Guardian online*. He asked me if I wanted to write a blog post, a quick seven hundred words, about why I was doing what I was doing and how my experience was going. This scared me: I was really busy and even though I welcome the chance to get my message out, my body and mind told me I needed a proper break, the kind in which you see nobody and do nothing. I asked Adam how many people would be likely to read the blog, to make sure it was worth my while. Adam said it would have at least a couple of thousand readers but if it went down well, tens of thousands were possible. That was good enough for me, especially as you never know where things might lead. What happened shocked both me and Adam. Within hours of my blog posting, an intense debate was raging. A few hours more, and it was climbing its way up to the top five of the website's 'Most Read' stories chart.

The *Guardian online*'s charts are something of a self-perpetuating system. People short of time go there to read the top news stories quickly. If a story rapidly rises to the top of this chart, it can stay there for days, moving out across the Internet.

And the intense debate sent my post straight to the top by the afternoon, with comments coming in thick and fast. About 60% of the comments were positive and supportive to an extent I had never experienced, about 10% were curious, and the remaining 30% thought that I was a middle-class 'trustafarian' (a trust-fund-supported, free-spirited, counter-cultural Bohemian) with nothing better to do.

Ironically, it was my critics who kept the debate going and the story at the top of the chart. The comments were totally polarised and the jury completely out. I had offers of marriage and casual sex (from both women and men). I had testimonials I didn't remotely deserve. I was a hypocrite for using a mobile and laptop. What I was doing was an insult to the poor people of Africa. I was a fame-seeking egomaniac on a publicity stunt ... My reality was in between: an ordinary guy, doing what he thinks is best right now, knowing, all too well, that he has as much chance of being wrong as of being right.

The post eventually became the most-read story on *Guardian online*, with 400,000 readers. Paul Kingsnorth (author of *Real England* and *One No, Many Yeses*) and George Monbiot (author of *Heat* and *The Age of Consent*, among much else), two people whose thoughts have greatly influenced me, joined in the debate. Adam was pleasantly surprised and asked me to do a follow-up post, whilst the *Guardian* itself wanted an article for their *G2* magazine. Off we went again.

I'd included a link to the Freeconomy Community website and it was going nuts. For days, it had a new member every minute. In one week, the community grew by more than 15%. During November, I got anywhere from 75 to 150 emails a day from well-wishers and people interested in bringing the online Freeconomy Community to its next logical conclusion: real life. Somehow, I even managed to get letters through the post, though I'd never disclosed where I was living. Interestingly, not one of the emails or letters was negative or threatening. The negative stuff seemed to need the anonymity of a blog comment, which my own blog had made me used to.

It was impossible to keep up with the emails and the world's media was back on my tail. One day, I did interviews with journalists in eight countries. It was insane: much too much for one person. I had a book to write with a deadline only six weeks away, a free feast and festival for thousands of people to organise, and the small matter of survival without money to think about. But it was an exciting time. To see the message I feel most passionate about spread across the world filled me full of joy and, thankfully, gave me a much-needed jolt of adrenaline. And for the first time in a while, I didn't feel so alone in what I was doing. A lot of the emails and letters came from people wanting to start their journey into moneyless living. Knowing that gave me a lot of strength for the weeks ahead.

14

THE END?

Never in my life had time gone by so quickly. The exhaustion and apprehension I'd felt on Buy Nothing Eve were fresh in my memory, yet the finishing line was in sight. Ten months ago, I'd thought November couldn't come quickly enough but now my only concern was the sobering thought of re-entering officialdom. It had been almost a year since I'd received envelopes with little clear plastic windows (a clear sign that it comes from a machine and not a person) in the post. I was getting attached to the idea of having no bank statements, utility bills or tax returns.

Barring major health problems, I thought nothing could prevent me from finishing my year. At this stage, I felt I could manage thirty days of almost anything. But I didn't realise that the final furlong would be the most mentally punishing of the whole twelve months. It wasn't that I struggled to survive; that was a piece of out-of-date cake. Rather, I had a unique opportunity to finish something I'd started, in a way that would maximise its impact and, I hoped, give birth to something bigger than just one man living without money. I felt there was more to life than staying within the confines of my comfort zone.

THE FREECONOMY FEASTIVAL 2009

A couple of days before the *Guardian online* blogs were published, my friends Francene and Andy had reminded me that I'd promised to end my year without money with a feast even bigger than the one that began it. I said 'no'. I had no idea that the story would cause such a stir on the website, and in the rest of the media, but nonetheless, I still said 'no'. I was barely treading water as it was and I knew the responsibility would, inevitably, fall on my shoulders.

But after a couple of days' thought and persuasion, I eventually agreed. Last year's feast had been one of the most uplifting days I could remember. I told myself I'd have plenty of time to sleep when I was dead. With high anxiety and trepidation, I agreed not only to organise a free three-course feast for hundreds of people but also a one-day festival of all things free. It was a great chance to show how Freeconomy could work, even in a city and if just for one day, and a fantastic way to celebrate the end of my year without money. It would be a huge challenge; making it free for everyone would depend on everyone freely offering whatever they could on the day. 'Let's make this massive', I thought. Last year's event had been a three-course feast for a hundred and fifty people. I felt that the growth and interest in the Freeconomy Community over the year meant it deserved to be a lot bigger this time.

I needed to decide what we wanted to happen on the day and make a list of everything we'd need for it to become reality. It felt like the list I made for myself at the beginning, except this was a list for just one day. But for thousands of people. I put a shout out to all the members of the Bristol and Bath Freeconomy Communities living within a twenty-five-mile radius of my caravan: in four weeks' time I wanted to put on the biggest moneyless festival the city had ever seen, with no cash, no funding and no monetary donations. I hoped for at least ten committed volunteers; any fewer and it would have been really difficult. I wasn't over-confident; the day was only weeks away. But the response was huge; a sign of how far the Freeconomy movement had come in twelve months. Even Brigit Strawbridge, star of the hit BBC series, *It's not Easy being Green*, got in touch, asking if she could volunteer on the day. Within a week, I had a team of almost sixty volunteers, most of whom I'd never met. Yet by the end, many of the volunteers had become friends.

I called a volunteers' meeting one evening, just three weeks before international Buy Nothing Day 2009. First, we needed to decide if it was possible to pull off such a mission in such a short time. Second, we had to decide what we were going to do, how we were going to do it and who was going to do what. After an extraordinarily efficient four-hour meeting, everyone had agreed that it was going ahead, that it was going to be big and that everyone had to be responsible for making a bit happen. This was Freeconomy in action.

NAPPY-FREE BABIES!

Parents interested in finding innovative ways to simultaneously save money and the world's resources often ask me about nappies. Disposable nappies are the norm in western society; few mums could, understandably, imagine anything different. They are, however, an ecological nightmare. According to the Women's Environmental Network some eight million are thrown away every day in the UK alone. British babies get through three billion a year. This costs their parents, on average, £500 a year; two full weeks' work for someone on a minimum wage.

Choosing laundered nappies could save this waste and cost less. But while terry nappies are an excellent alternative, there are other options. 'Nappy-free baby' (www. nappyfreebaby.co.uk) or 'Elimination Communication', as it is also known, is a potty-training method in which the parent or care-giver uses signals, cues and intuition to deal with the child's need to poo. The ideal is to use no nappies at all but in some situations they may be needed. Elimination Communication not only drastically cuts down the world's nappy mountains, it also empowers parents to become more attuned to their kids.

The practice was inspired by the traditional methods of less industrialised cultures, so whilst it seems new to most of us it's merely a revival of ancient knowledge.

I've seen it work and I was absolutely shocked. I had no idea that a baby didn't need a nappy and I was even more shocked that I hadn't realised it long ago!

There were just two tiny hurdles. The first was promotion: this was needed straight away but couldn't happen until we'd overcome the second hurdle: finding the venue. This event was going to be big. We needed to find a large place, in a central location, where we could have the space for free. No small task.

Francene, responsible for talking me into what became an intense and highly-pressured 180 hours of work at exactly the same time the world's media once more picked up on my story, got her rear end into action in late, but typically fine, form. She got in touch with Oli Wells, director of an up-and-coming venue in Stokes Croft, an area of Bristol that, in the last three years, had been transformed from a place rife with homelessness and drug abuse into the artistic quarter of the city. She explained what we wanted to do and why. Oli was incredibly enthusiastic and offered us the entire second floor of his highly sought-after venue for free. He said if we could make a DVD of the whole event and its run-up he'd appreciate it but, in pure Freeconomic style, he didn't make it a condition.

There was one slight problem with the venue; it didn't have a kitchen. Actually, it didn't even have running water. We had the perfect space but we had to find cookers, gas, cutlery, tables, chairs, boilers, utensils, plates, glasses and everything else that a restaurant needs. We'd have to borrow everything, get it to the venue for one day and then get it all back – in one piece – to the right people the next day. I tried not to let it overwhelm me but this really was a monumental moneyless mission.

I made a list, together with our head chef, Andy Drummond. I sent a few emails and spoke to a few people. Within a week we had been promised, by seven organisations, enough kitchen gear to feed and seat about a thousand mouths for an evening. (One of the organisations was the charity I'd made a cash flow forecast for earlier in the year.) Volunteers who'd been assigned the task of finding the gas we'd need to power the cookers came up with some seventy pounds of butane otherwise destined to sit in sheds, never to be used.

We needed to get all the stuff there in time. A team of drivers and cyclists got together to make that happen. Next on the list was food; only second in priority because if we'd had neither venue nor kitchen, we wouldn't have needed it. We set up three wild food foraging teams, one led by Fergus, another by Andy and the last by James, who also volunteered to teach his crew how to gather and press apples for juice for the day.

Three skipping teams, led by Cai and Abby, our most experienced skippers, were set up. Cai could climb fifteen-foot-high walls and slide down lampposts in a way that made me believe he'd done a stint either as a fireman or in the circus. Our idea was to combine getting the food for the feast with teaching people how to forage and skip; an entirely mutually beneficial set-up. I'm a big believer in learning through doing and this was certainly that.

The whole event was about education and skill-sharing. Some of the people cooking on the day were trained chefs, others had never cooked for anywhere near that many people and some had barely ever cooked for themselves. It was my first effort at building a kitchen from scratch and I learned a lot in a very short time. Everything from gas regulations to the logistics of – without spending a penny in the process – getting an entire kitchen delivered to a venue one day and sent back the next.

While the foragers and skippers were busy finding food from the wilds and the bins, I linked up with local food businesses and organisations. I went to meet Pete and Jacqui, two organisers from the Bristol branch of Fareshares, to see if they wanted to get involved. I had a huge amount of admiration for the work they were doing. Just like everyone else we spoke to, Pete and Jacqui said a very enthusiastic 'yes'. Fareshares has set up formal links with some supermarkets; whenever the supermarkets know they won't be able to sell some of their food, Fareshares picks it up and delivers it to places, like homeless people's shelters, which probably wouldn't be able to survive without it. But sometimes, even they had too much waste food to get rid of. They were happy to help and I promised to mention them in any interviews I did on the day. From them, we got everything from bread (some two hundred organic loaves), beans and Bombay mix for snacks, to the loan of three hundred glasses. Their contribution eventually became a large van full of a couple of tons of food; enough to supply the basics of the meal.

I set up links with a few local wholesalers, who also suffered the frustration of being bound, by law, to throw good food in the bin. A local organic food co-operative, Essential, supplied a few types of food that we couldn't get from Fareshares: couscous, bulgur wheat, rice, flour, nachos, rice and soya milk, crisps, chocolate and huge bags of other snacks. Either we were in luck or our current food system is wasteful in the extreme. Experience told me it was the latter.

One crucial ingredient was missing: alcohol. But Andy Hamilton, and a team of merry home-brewers, was ready. Three weeks before Buy Nothing Day 2009, they'd set to, brewing up about ninety gallons of beer. They made everything from molasses beer and yarrow ale to a spicy brew containing ingredients, such as cinnamon, that I would never have imagined working in an alcoholic drink. But they did; all the brews came out fantastically well. A couple of gallons of spirits were donated from people's drinks' cabinets. We officially had a party going on.

Francene, Fergus, Cai and I went out the night before the Feastival. After walking down a few dead ends, we struck gold in one skip: seven hundred jars of organic fair-trade chocolate spread that would have cost £2000 in the shops a few weeks earlier. Being mostly sugar, realistically, it would still have been fine to eat in five years' time. But the law is the law and doesn't allow for much human discretion.

The three-course feast – indeed all the food on the day – would be totally vegan. But we were missing a vital ingredient: fruit and vegetables. I got some from Christina, from Somerset Organic Links, an organic farmers' co-operative that pools their harvests and resources in a successful attempt to halt the large supermarkets' control of the food industry. Christina eventually supplied some two hundred pounds of vegetables; fantastic but about three hundred pounds short of what we needed.

Abby (an American who'd moved to the UK with ideas of living without money) led a team of bin-raiders for a couple of nights, coming back with vegetable bounties galore. But even that wasn't anywhere near enough to meet the demand we anticipated. We organised a team, Elly, Fergus and Cai, to go to the local fruit and vegetable market, from which about fifteen wholesalers operated. It was a risky strategy. They couldn't go until the morning of the Feastival; if they went any earlier, the wholesalers wouldn't have had their final orders or know what was about to go off. But, deciding it was our best bet, the team went along anyway. I was pretty nervous; thousands of people were expecting food. And just a couple of hours before the morning team of twenty-five expected to start preparing and cooking the vegetables, we had only half of what we needed.

This was the morning of my last 'official day without money'. I stayed behind for interviews, whilst the final skipping crew went off, fingers firmly crossed. I tried to focus on plugging the Feastival and the website in every interview I could, but my mind couldn't help wandering to the wholesale market, wondering whether the team was being escorted off the premises empty-handed. In the middle of an interview for BBC Radio Kent, Fergus's local station, I got a text message from Cai: the eagle had landed and they had a van full of food. The guys at the market were happy to help: they also hated the weekly routine of chucking out good vegetables every Saturday morning. The Freeconomy Feastival was on!

Food was only one part of the day. Elsie and Katey, two volunteers from Stroud (a small town north of Bristol), spent two weeks gathering clothes for a massive free clothes shop and swap, to which anyone could come, leave things they were bored with, or take stuff they fancied. They also made a creative corner to show people how to mend clothes and make useful things from stuff like old packaging. Julia, Elly and Di gathered books for a book shop and swap, collecting hundreds of books before the day even began. I organised a day-long programme of eight talks, including people like Claire Milne (one of the food policy advisors for Transition Towns), Alf Montagu (regular contributor to television programmes on freeganism), Ciaran Mundy (an alternative economics advisor for Transition Towns), Fergus (on making all sorts of ridiculous stuff from wild food) and me, sharing my experiences of living without cash for a year.

Sarah volunteered to organise the day's entertainment, collecting some of the best-loved musicians on the Bristol scene. These bands would usually cost a tidy sum to hire for the night, even if you're lucky and they were available on a Saturday evening. She didn't have to ask any of them; they contacted her, offering to perform for free. They seemed just as enthused as we were. Not content with that, Sarah also managed to get a pedal-powered stage from a local project, Bicyclette, which meant the entire night's music would be off-grid. People from the crowd took fifteen-minute stints on the bike to keep the amplification going. I'd managed to get a smoothie bike – a pedal-powered smoothie maker – from the Bristol Food Hub. This kit would normally cost anywhere between £150 and £250 for a day's hire but both organisations offered it for free. It seems that when you start something with the intention of giving and not taking, it is almost impossible to stop others wanting to do the same.

I organised a free cinema where we showed movies with different themes: 'Money as Debt', 'The Story of Stuff', 'Earthlings', 'The Age of Stupid' and 'The Transition Movie'. We also had an incredibly funny, yet enlightening, stand-up comedy piece by Rob Newman: *The History of Oil*. Some alternative health practitioners, from a few floors up the building, who normally charged over £30 an hour for acupuncture, massage and other therapies, offered their services for free. I saw the therapists later, enjoying the food and ale and dancing to the music. This reinforced my belief that there really was another way of doing things, a way based on giving rather than exchange. A way that could really work.

After three full days of getting the venue set up and the food sorted, the day was upon us. I had a pretty busy schedule, including a series of sixteen interviews, starting at six o'clock in the morning and going on through the day. One of these was a live interview on BBC News 24, which was partially responsible for the massive queues that formed outside Hamilton House from the moment the doors opened. On top of the interviews, I had the small matter of a fourteen-hour free feastival to organise, including giving a ninety-minute talk in the middle.

It sounds worse than a date with Maggie Thatcher. But it was one of the most fulfilling days you could possibly imagine. The atmosphere – both in the kitchen and the crowd – was incredibly positive and uplifting. The thousands who came really could not understand how it was all for free, with no donations or funding allowed. People from many different socio-economic backgrounds (I saw business people and homeless folk talking as people, not labels) came together to enjoy a rare day in which everything they could imagine was totally free.

By seven in the evening, my only remaining task was to sit back, eat the amazing food that the kitchen team had prepared, including Fergus's fantastic sea beet sorbet and various curry and pasta dishes, and drink good home-brewed organic ale whilst listening to some of my favourite bands. Four weeks of intense work had paid off. We fed almost one thousand people at least one dish each, and over three and a half thousand people came through the door. People talked about it for weeks afterwards, unable to believe it really was all done without spending a penny. Freeconomy, it seemed, was definitely no longer the domain of greenies, lefties and hippies.

It was an emotional day. Seeing everyone giving whatever they could to the day, with no thought of anything in return, was immensely inspiring. To me, it was the most beautiful example of how things could be if we chose to live life thinking 'how much can I give?' rather than 'how much can I get?' Some of the volunteers relentlessly dished up food for twelve hours with barely a break. How many paid staff would do that? But they had genuine smiles on their faces.

And as much as it was really hard work, with little but the joy of doing it in return, we were all sorry when it was over. I met so many amazing people and made so many new friends. The whole experience created a fantastic bond.

For the last few weeks, my head had been over-ruling my heart, telling me to go back to living with money. This was partly because of complications in my long-term vision for the Freeconomy project and partly because I felt I needed a break. Living without money wasn't as difficult as I'd first imagined but doing it in a society driven only by the desire for more felt like I was swimming against a strong tide. But I found the Feastival so inspiring that I decided to delay making a definite decision. My emotions were running high and I felt I needed to let things settle. Whatever way I went, it would be a major life decision.

TO CONTINUE OR NOT TO CONTINUE

Life had been insanely busy for two months. I hadn't had a proper chance to think whether I wanted to continue living without money once my year had officially ended. In some respects it was a simple decision, yet, right up until the last day, I was torn. My heart – and many parts of my head – said a huge 'yes'. I had never felt happier, healthier or fitter in my life; why go back to a less enjoyable way?

However, life is rarely that black and white. I had finalised a book deal a few weeks earlier, which meant that money was waiting for me. The book would be sold for money no matter what I decided to do. And it would generate royalties that I could use however I wanted. I had to choose what I would do with the proceeds:

1. Let the publisher keep the money. This wouldn't have appealed to my agent, Sallyanne! She had been fantastic to me all year, allowing me to refuse fees that both she and I would

normally have received. And she'd put in a huge amount of work editing the book.

2. Let my agent keep all the proceeds. I'm sure Sallyanne would have been very happy!
3. Give the proceeds to a project I wanted to support.
4. Set up a trust fund to help buy a piece of land for the first 'real' Freeconomy Community. If I chose this option, I'd decided I wouldn't own the land and the community would be run by its members through consensus.

I had no idea what to do, so I posted a blog on the Freeconomy Community website to ask for advice. The response was one of the biggest ever; more than five hundred people either commented or emailed me.

THE DECISION

The result was overwhelming. About 95% urged me to take Option 4. (Maybe they wanted somewhere to come and stay every now and then for free!) I made the big decision: go with the majority opinion of Freeconomy members and remain moneyless for as long as I could in the mean time.

I came in for some mild, though very well-intentioned, criticism from the 5% who wanted me to choose Option 3. This criticism was hard to take, coming from people I respected and, more importantly, almost entirely agreed with. At heart, they, like me, were idealists. However, over the years, I've learned to let the idealist in me have regular conversations with the realist. Two years ago, without doubt, I would have chosen Option 3. Was I getting wiser or falling off my path?

The critics said the real Freeconomy Community would no longer be moneyless if I bought the land. It could offer no solution to society, they said, and would be a farce. I couldn't fully disagree with them. But life, I guess, is full of such dilemmas; all we can do is chose the best option, give it a good shot and question its rationale every day. These critics didn't know that I'd paid for the Freeconomy Community website, the infrastructure they were using for their comments, from the proceeds of selling my houseboat. Does the fact that I paid for the website negate the fact that it enables thousands of people to move towards a more moneyless life, and the re-building of resilient communities? Or is the work it is doing right now all that is important? I felt the two situations had important parallels to be examined. And there was something else I felt was relevant. Slaves often had to buy their freedom so they and their children could be free. Is it acceptable to make a one-off payment to buy one's long-term freedom? Or does paying the slave-master reinforce the system you want to change? I still don't fully know.

FREE PERIODS

When you start to live without money, the first problems you need to solve are those areas where you use disposable products. Obviously, you can't buy them. And disposables consume both time and resources.

Being a man, the question of moneyless menstruation is a tricky one. Women's health is certainly not my forté. For coping with periods, most women choose disposable sanitary towels. According to the waste consultants, Franklin Associates, in 1998 6.5 billion tampons and 13.5 billion sanitary pads, plus their packaging, ended up in landfills or sewer systems. For coping with periods without money, there is an obvious solution that even I know about: a mooncup. This is a rubber cup, which the user inserts in her vagina to collect the menstrual flow. It's held in place over the cervix by suction. Looked after, a mooncup can last a lifetime, enabling you to use less money and really help the environment into the bargain.

Again, the option that saves money is also the option that could well keep our natural environment habitable for humans.

THE FREECONOMY COMMUNITY LONG-TERM VISION

I chose Option 4. I decided to set up a trust fund to which all the proceeds from this book will go. The money will go towards buying the first piece of land where this project can put down roots. At the time of writing, the fine details have yet to be thrashed out: the year itself consumed all my time and the few months since it ended have been entirely focused on writing this book. But I have the vision in my head.

The community will be based on the same principles as the online Freeconomy Community and my year without money. We'll put the infrastructure in place using as little money as possible and as much local and waste material, human passion and determination as we can. There will be a transitional period after which money, whether notes, coins, cheques or e-money, will not be used. It will be a community with food, friendship, fun, fire, foraging, music, education, resource-sharing, dance, art, care, skill-sharing, experience, respect and scavenging at its core.

In Permaculture terms, we aim to be a 'closed loop system', in which we meet our needs from the local environment. However, in terms of inclusivity and outreach, as far as the land is able to support us, we intend to be the most open community we possibly can. Every member of the online community will be welcome to come and get involved. And when they leave, they will be welcome to take away any ideas they've found to be useful and incorporate them into their life. But it won't stop there. The community will be open to everyone who needs it and to those who want to spend a little time exploring moneyless living as an option for their future.

We will mix low-impact living with high-impact education and experience. I believe in education through doing, so much of the learning will come through living everyday life. We will go out on the land with people who know what they are doing and, in the process of helping each other live, everyone will learn what they want and need to learn. I intend the community to become a centre of excellence in sustainability, with teaching by the world's top practitioners. The teachers will give their time and share their skills for free, we will provide the land for free and the students will learn for free. Hopefully, they will then pass their learning on to others for free.

This is exactly how Freeskilling works. Freeskilling has now got to the stage where we don't have to look for great teachers; they offer to share their skills and we accept gratefully. Sustainability courses often cost too much for volunteers and people on low wages. This will certainly not be the case in the community. I want to see people from all walks of life enrolling, not just those far down the path of more ecological living. Education really can be free. All it needs is the determination of those who can help educate others.

Skill-sharing will be part of life for the people who live permanently in the community. The core group of diversely skilled people living there will share their skills over time. One day, the carpenter will help the forager; the next, the forager will help the grower. One evening you'll be out collecting waste; the next, you'll be cooking dinner for those who are working hard on other tasks. Everyone will be able to find out what they really love doing, with the flexibility to do something different if they want to. If someone crucial to the success of the project needs to leave, any number of people will be able to help fill the void until the next suitable person comes along.

It won't be easy to find the perfect piece of land but I hope it will be within a fifty-mile radius of Bristol. To be self-sustaining, you need resources. Most important is a source of water, ideally a river, not only to provide drinking water, but also for easy washing, energy creation through micro-hydroelectric power generation and, not least, summer swims. Some woodland will be crucial: some established fruit trees would bring us even closer to the perfect plot. But I'm not holding out for perfect; the likelihood is that if it satisfies some of the criteria, it will be viable.

The community will be a kind of sustainability 'theme park'. It will include every type of low-impact dwelling for which we can get permission: earthships (I'd love the farmhouse to be an earthship), other types of passive solar home, rammed-earth structures and more. It will have a reed-bed waste system, compost toilets to make humanure, forest gardens, greenhouses, beehives, wind turbines, cob ovens and rocket stoves. The key will be in the design. If we get that right, we'll almost completely eliminate waste production. The land will work with us, and we with the land, to make the community as energy-efficient as possible.

It's not clear what the community's legal structure will be. I do know that, at the beginning, until it can stand on its own two feet, there will be a sort of steering committee. This will guide the community through its infancy and hold it true to its original intentions and integrity. Just like a parent looks after their new-born child, the committee won't own the child but they will help it through its formative years. There will be a set of core guidelines – like being moneyless and organic – from which the community may not stray but other than that, its structure will be created by the people who live there.

There are many obstacles to this vision: tax, planning permission, social pressures, local opinions and the small issue of acquiring a suitable piece of land. And that's just a few of the more obvious ones. All these issues will have to be tackled sometime. And if not by us, then by whom? And if not now, then when? Should we leave it as a battle for the next generation to pick up? After all, it'll affect them more than us. Or should we, as parents, try to ensure that our children inherit a nice habitable planet when our time is up, in the same way we would like our kids to inherit a nice house that we've worked hard, all our lives, to pay for?

BETWEEN THE DREAM AND THE REALITY

My year of living without money officially ended at midnight on Sunday 29th November 2009. I'd done it. I had an obvious get-out clause, if I wanted one; I'd completed what I'd set out to achieve. But I didn't want to get out; I really wanted to keep going.

Making the decision not to go back felt like I'd lifted a heavy weight from my shoulders. And the support I received from my friends and family was huge. They didn't see it as strange; I felt they had accepted my choice not because they loved me, or despite loving me, but because they could see how the experiment had worked and how happy it had made me.

Almost immediately after I'd made my decision, I knew it was the right one. A couple of days after the Feastival, I walked through Bristol's main shopping centre and I took some time out to observe what was going on. I felt as though people had lost their minds. In the US in 2008, a supermarket employee was killed when a stampede of bargain-hungry consumers could no longer be held back from the start of the sale, trampling the guy to death in their rush to the aisles. A similar situation happened here in the UK in 2005, at the opening of a giant furniture store. Several people were crushed (not fatally) by others looking for the opening event's bargains. In Saudi Arabia in 2004, three people were killed and sixteen injured in the name of a bargain 'hunt'. How far have we gone when we trample someone to death to save a few quid?

It was the height of the Christmas shopping season and the shopping centre was mayhem. Through the bustling shopping crowds came a group of people holding a sign: 'Free Hugs'. For fifteen minutes, they did exactly that: they gave a free hug to anyone who wanted one. A queue built up, such was the popularity of their 'product'. But free hugs don't make money; they were quickly escorted out by security guards. The shopping centre looks like the public street but the land is privately owned; they weren't allowed to give away even hugs for free on corporate land. It seemed to me that in today's consumerist culture, you're allowed (indeed, actively encouraged) to consume far more of the Earth's resources then anyone could actually 'need' – but don't try and hug somebody on the way.

Despite such reminders that I live in a world driven by an addiction to accumulating more and more lucre, my year without money had given me a huge amount of hope. Every day, I'd get countless emails and blog comments from people who said that, while they couldn't see themselves going completely moneyless, they really wanted to make big changes in their lives. Some wanted to 'downsize' and cut their consumption, so that they could work less and live more. Many wanted to reduce their carbon footprint drastically. Others just wanted to start recycling their waste. Even more encouragingly, hundreds wanted to come and help create the first moneyless community in contemporary society.

We're a long, long way from living sustainably, let alone living without money. But more and more people are aware of the future challenges facing humanity. Every year, more and more column inches in newspapers and magazines are devoted to environmental issues, and climate change stays at the top of the news. People really are making changes: some small, some huge but in a more ecological direction. I know it will take time. But it's vital to plant as many seeds as we can now, if we want our children to benefit from the fruit. Just because you won't get to sit under the shade of the oak tree doesn't mean you shouldn't plant the acorn.

I got off the bench and walked north, out of the shopping centre, and looked back and smiled. Whatever happens – whether we embrace change or consume ourselves into oblivion – it is important to remember that, in the words of the legendary comedian Bill Hicks, 'it's just a ride'. Enjoy this gift for what it is, not for what you want it to be.

15

LESSONS FROM A MONEYLESS YEAR

No matter what way of life you choose, lessons appear every day. The problem is, we're not usually very receptive to them. Worse still, we often see the lessons as failures, hassles or even disasters, rather than as a chance to learn something new. In *The Road Less Traveled*, M. Scott Peck said: 'Life is difficult ... but once we truly understand and accept this ... then life is no longer difficult'. In some respects my year was difficult and in others it was the happiest time of my life. In the summer of my experiment, I'd accepted that life isn't always meant to be 'perfect' and that I had no god-given right to everything this society tells me I could have. I surrendered to the fact that life was just the way it's meant to be at all times: perfectly imperfect. After that, accepting the little hassles, the little inconveniences that living without money inevitably throws your way became fun.

My experiment was a complete change in how I lived. I learned more things in that year than in any twelve-month period I'd ever lived through. Some so subconsciously that I didn't even know that I learned them.

DON'T UNDERESTIMATE OTHERS

One of the hardest things about moneyless living was the thought of what other people might think. I wasn't so bothered about society in general but I was worried that my parents would think I was throwing away everything I'd worked hard for. This concern turned out to be completely unfounded: the aspect of the year that I have felt happiest about was my parents' reaction. I'm not sure what they thought about it at the beginning; we didn't talk about it much. I'm lucky: even if they had disagreed with my stance – and they may well have – they'd have given me whatever support they could. It may have been hard for them to take at first. They'd watched me work thirty hours a week for four years to pay my way through my degree, and they'd helped me out a lot along the way. Now they watched me renounce it all.

It's been interesting for me to watch them go on their journey since I started along my path. At the beginning, I'd ranted on, telling them how everything they were doing was wrong, how my opinion was right and how they needed to change. Understandably, this erected walls, defences through which none of us could properly communicate. But it was I who needed to change. What made my opinion more correct than theirs – or anyone else's for that matter? I stopped my pestering. It seems children's pester-power only works if they are trying to get their parents to buy more, not less.

About six months after my decision to leave them in peace, I noticed small changes. One time, my mum phoned to tell me she and dad had decided to become vegetarian. Another, she rang to say that she was going to stop buying so much stuff. Just by me providing information, with no judgement or claim to rightness, my folks started to question things themselves. Not because I was telling them to, but because they wanted to. Eventually, they got right behind my moneyless plans and life. There's no sign they're going to join me on the path but they are constantly questioning how they live their lives and are making little changes almost weekly. They've offered to help in any way they can with the setting up of the community. I don't expect them to live like me, just as they don't expect me to live like them. They've given me lessons about what it takes for us to co-exist on this planet.

I would never recommend not standing up for what you believe because of what other people might think. But I am beginning to realise I have no right to criticise others for flaws we all have, or have had. It is much more constructive to support each other in making even the smallest change that is positive for the whole planet. This way, walls get demolished, and we can have a proper dialogue.

A HALF-WAY HOUSE

I would love to live in a moneyless world. No doubt about it, that is my ideal. But whilst I will work and move in this world as if that were a real possibility, the realist in me knows it isn't going to happen, at least not in my lifetime. The overwhelming majority of people have no desire to give up money: they think it is a very useful tool. And many of those who would like to give up money have told me, repeatedly, they don't believe they could.

The support I received over the year, both from the media and the public, has given me so much hope for the future. I truly believe that we can make the changes that the world's ecologists believe we need to make. One change I believe we can make, one that is realistic, if not imperative, is to move to local currencies. A local currency operates purely within a town, village or small area. In the UK, examples include the Totnes and Lewes Pounds but there are examples in other countries. Local currencies aren't legal tender, more a kind of formalised barter, in which produce or skills are traded for an agreed amount of local currency, which the receiver can then 'spend'. Local currencies aim to keep 'money' circulating in a community, build relationships between producers and consumers, get people thinking about where and how they spend the currency and encourage local businesses and trading. Whilst users of local currencies must still partake, to varying degrees, in the global economy, local currencies are a huge step towards re-localisation of economies.

Local currency is based on exchange and therefore doesn't have some of the deeper benefits that I believe pay-it-forward economics could have, but to me, it's a good half-way house. Local currencies are a fantastic method of reducing the degrees of separation between the consumer and the consumed; users of local currencies have a much fuller appreciation of the processes of production and whether the producers' needs are being met. If some communities could make a complete transition from the current monetary system, this would be a sustainable model of living that other communities could copy.

COMMUNAL-SUFFICIENCY

When people learn that I live without money, most assume that I must be almost completely 'self-sufficient'. That was my plan, but I quickly learned that independence is one of the biggest myths in modern society. At the very least, we depend on bees, earthworms and micro-organisms just to survive. Not only did I realise that I couldn't become completely self-sufficient even if I had wanted to, I also realised that I had no desire to be; some of the greatest happiness in my life comes from the relationships I have with people in my community. What I believe works best – and what I find most desirable – is for small numbers of people to work interdependently, together building 'communal-sufficiency'.

Robin Dunbar, the British evolutionary biologist, has studied the tribe size of non-human primates, from which he has developed his description of the 'Dunbar Number'. He estimates that humans can maintain stable social relationships with approximately 150 people. These communities can be streets, suburbs or villages. Around this size, I believe communities can benefit from the economies of scale that come into play as we produce things for larger and larger numbers, without causing the ecologies-of-industrialisation that arise when that scale becomes so large that it becomes inherently unsustainable. Because I lived my year in relative isolation, I had to do most things myself. To cook my dinner, I needed to gather and chop the wood, gather and chop the food, feed the rocket stove for thirty minutes, serve up, and wash the dishes. If this had been an interdependent process, I would only have had to do one or two parts, giving me time to relax or do something creative. The beautiful thing is that you don't need money when you live within a community – you bring what you can; your reputation, in a way, becomes your currency. The more you give, the more you'll find that you receive. That has been my experience, anyway.

THE ESSENTIAL SKILLS OF THE FUTURE

Before I started my year I believed that the most important skills I'd need to live ecologically and without money would be things like carpentry, vegetable-growing, permaculture design, medicine, clothes-making and repairing, cooking, bushcraft and teaching. I still believe they are absolutely essential to moneyless living, especially if we want to create a self-sustaining community. However, I would now call these 'secondary skills'. I believe that physical fitness, self-discipline, genuine care and respect for the planet and the species that live on it, and the ability to give and share, are the 'primary skills' for this way of living. Without at least some of these skills, it is not a way of life you can either embark on or will be able to sustain. At the community level, it's not quite so important that everyone is physically fit; many jobs don't need it. And if somebody becomes ill, there are others around to help. But the healthier and fitter the individuals involved are, the better. They'll enjoy it a lot more, as a lot of the fun stuff involves being active outdoors.

I cannot over-emphasise how un-skilled a human being I am. I am ordinary beyond belief. But if I can live this way, many people also could, if they really wanted to; and most people would probably be a lot better at it than I. As long as the will to do it is in there somewhere, the rest is a matter of education and practice. It's much easier to teach somebody how to plant a seed than to convince them of the need to plant it.

AN ORGANIC FLOW OF GIVING AND RECEIVING

From the moment we are born, most of us are taught that money, not community, is our primary source of personal security. It's perfectly understandable that most people have taken to protecting what they already have; otherwise, if things go badly, what are they going to fall back on?

One of the first, and most important, lessons I was taught by moneyless living was to trust life. I firmly believe that if we live each day in the spirit of giving, we'll receive whatever we need whenever we need it. I've long since stopped trying to explain this intellectually; it comes from feeling and from life's experience. Getting a free caravan after selling my houseboat to pay for the Freeconomy website was a major example, but many little things happened daily. On many evenings, I'd cycle from house to house on my way home from the city, dropping off food – to friends and people who needed it – that I couldn't eat myself. Other evenings, I'd find myself in the city, hungry after the cycle in but having forgotten to take food with me, only to meet a friend or acquaintance in the street who'd invite me for dinner.

My experience has been that when you give freely, with no thought of what you'll get in return, you receive freely, without fail. It's an organic flow of giving and receiving, a magical dance that our entire ecosystem is based on. But it requires a leap of faith, and placing trust in nature to provide for your needs. Christians call it 'reaping what you sow', Buddhists call it 'karma' and atheists call it 'common sense'.

Take this for an example: let's put ourselves in a group of thirty friends. We decide that we are going to be aware of each others' needs and do our best to meet them if we can. Each person in the group now has thirty people looking after their best interests. However, if each of us decided to go back to living as most of us do today, thinking mostly about ourselves, we'd only have one person looking after our best interests – ourselves.

If we put a bit more love, respect and care into the world, I believe we will all benefit from a world with more love, care and respect in it. It's not a complicated theory. Staying in the flow of giving and receiving freely is a challenge. I don't always succeed. But the times when I'm in that flow are my happiest. Life seems easy, there is no resistance, no swimming against the tide. Trusting in life to supply whatever you need is, for me, complete liberation. It frees you from worry and enables you to do whatever it is you really want to do.

MONEY IS JUST ONE WAY OF DOING THINGS

Throughout my year, many people suggested I could only live without money because others live with it. 'How would you have a road to cycle on if there weren't money and I didn't pay my taxes?' It's an understandable argument, but it's based on the underlying assumption that you need money to create things. An assumption that is, I believe, fundamentally flawed.

I have come to learn, more and more, that using money is just one way of doing things. It is a way of apportioning reward to those who help build the road but it is completely unnecessary for the road's construction. Money allows you to use labour that isn't local; the road's asphalt will almost always be made by people far away. Living moneyless forces us to obtain the materials we need locally; it forces us to take responsibility for meeting our community's needs; it forces us to have more appreciation for what we use. It also forces us to use local labour; something I feel is absolutely vital to successfully tackling critical issues such as peak oil and climate change. There is no reason why local people cannot build whatever roads and paths they need. If we devolved decision-making to communities, what would there be to stop local people coming together to create whatever they need? Nothing more than a change in perspective.

I've been criticised for using my bike on roads, especially in interviews. I understand this; it looks like there's an inherent hypocrisy. But you can't gouge out a man's eyes and then criticise him for being blind. I have to deal with the world I'm in, not an ideal world that doesn't exist. I don't want to maintain this world but it is where I am. A bike is my way of finding balance between having as much impact through positive social change as I can and as little impact on our natural environment as I can. If it were up to me, I'd happily sacrifice large asphalt-covered roads if it meant we could get back to a truly sustainable way of living. And you can apply the same arguments to anything we want to create, whether it be houses, bridges, hospitals or schools. The more I live this way, the more I know another, more localised, way is possible.

NECESSITY IS THE MOTHER OF INVENTION

I knew before I began my year that I could only plan for so much. The vast majority of stuff, I felt, I'd have to work out day by day. It's an old adage but necessity truly is the mother of invention.

I didn't learn the wild fennel and cuttlefish bone toothpaste trick until a month into the experiment; the thought of horrifically bad breath forced me to investigate my options. I didn't use my hand-turned vintage Singer sewing machine until the crotches went in two pairs of jeans. I'd never heard of the concept of unpunctureable tyres until I wondered what I was going to do if I got a lot of punctures. I had no idea how to change my bike's brake pads without money, until I realised that bike shops throw out half-used ones and there were people in the local Freeconomy Community who could show me how to fix them.

The experience of my year has given me a lot of hope. Environmentalists pose post-peak oil apocalyptic scenarios about how it is all going to end really badly. I can understand the fear and scepticism; I share it sometimes. I agree that we need to start making the transition to re-design society for a time when our meteorological and economic climates are not so stable. If we can start making that transition now, I know that we will be able to work on whatever is thrown our way. Humans are an incredibly resourceful species; when things have been tough, we have worked together to come up with solutions. During the Second World War, the British worked together in the 'Dig for Victory' campaign. Whilst times were different then, when people knew their neighbours and fellow townspeople, and communities were smaller, I know that if we set about re-building resilient communities today, by re-connecting with the people in our local area, we would be able to cope with whatever the future holds.

THE REAL VALUE OF THINGS

Massive factories, supermarkets, hyper-stores and their ilk have completely changed our perception of the fair price of things. I notice this acutely whenever I work in the little organic food co-op in Bristol. People who say there is no way in hell they are paying £1.50 for a pound of courgettes have no experience of the amount of effort needed to grow them organically without huge fossil fuel inputs. A grower working on a minimum wage, as the vast majority do, has about five minutes to do all the work required to break even on that pound of courgettes. On average, the grower receives only about half of the retail price. From that half, they must apportion part for overheads and other direct costs. How much more quickly can we expect growers to work, if they are using their hands rather than energy-intensive machinery?

The more I have become responsible for producing my own stuff, or at least got closer to those who produce it, the more I realise the real value of things. My mate Josh makes magnificent chairs from willow he himself grows. I know how long it takes him to do it, from planting the setts to fixing the rods together. I know the real value of that chair and it goes beyond money. For Josh, it symbolises his respect for the earth and represents everything he stands for.

I've come to realise that large organisations who offer low prices only do so because they exploit people and benefit from economies of scale. Are they eventually going to strip this planet bare of every natural resource? Do their prices include the cost of the destruction of everything that we have been given? What price would their products have if they did?

FINAL THOUGHTS

We're at a crucial point in history. We cannot have fast cars, computers the size of credit cards, and modern conveniences, whilst simultaneously having clean air, abundant rainforests, fresh drinking water and a stable climate. This generation can have one or the other but not both. Humanity must make a choice. Both have an opportunity cost. Gadgetry or nature? Pick the wrong one and the next generation may have neither.

EPILOGUE

Learning to live without money – to change the mentality and habits that you've formed throughout your life – isn't something you can do, or probably would want to do, overnight. For me, it started seven years ago, when I read the book about Mahatma Gandhi and began what I believe will be a life-long endeavour to put his philosophies – mixed with my own – into practice in a modern context.

Since I started the Freeconomy movement in 2007, I've been looking for ways to take money out of the equation in all aspects of my life, from how I get my food, to how I have fun, to how I get from A to B. I've been looking for ways to replace money with real relationships with the people of my local community and with the natural environment. This takes a lot of time. Much of the information I needed came through experience and through meeting the right people at the right time. That's one thing I have noticed; the further down this path I go, the more people who are striving to live the same way enter my life. I am not sure if they were always there and I've only recently become aware of them, or whether the idea of living without money, an idea as old as the hills, is becoming more relevant as critical issues like climate change, banking crises, peak oil, environmental destruction and resource-depletion emerge. Who knows? It's clear that, for a plethora of reasons, moneyless living is a movement whose time has come and one that is growing rapidly.

Walking the path to moneyless living is like walking into a virgin forest at midnight without a lantern. You sense it might be a fantastic place to live but it seems daunting; sometimes overwhelmingly daunting. You have no idea what lies ahead or how far you must walk. Nonetheless, you walk. Inevitably, you stumble, fall over, hurt yourself, but get back up. A few hours in, you meet a stranger, trying to reach the same place by another path. You help each other. That someone else wants to find the same place as you not only makes you feel more physically secure; it also reinforces your belief that this is a place worth going to. You feel less alone and more sane. At four in the morning, as the darkest part of the night is losing its stranglehold on your perception, you see a group of people ahead, all looking for the same place. You join up and walk with them. You note landmarks, jot down directions and hang flags as a guide for others who may want to explore the forest themselves.

The closer it gets to sunrise, the more people you meet and the less frightening the forest becomes. The wild monsters you'd feared hadn't materialised. Suddenly, you reach a small clearing. It looks as though someone lived there many generations ago. You, and all the people you met along the way meet, at exactly the same moment, lots of other people coming to this place from totally different directions. Like you, they're looking for what their intuition told them might be paradise. As every seeker converges on this one spot, the sun rises over the clearing's horizon. Its light shows this place to be as magnificent as everyone ever imagined it would. There is abundance. Everyone helps each other pick fruits and nuts and shares their harvests. People build shelters together and there is more than enough for everyone's needs. How everyone managed to converge at the same time, from different directions, without a map, is one of life's mysteries. Some didn't even know why they walked the path into this forest. They just knew the path they'd been walking wasn't as enjoyable as they'd first thought it would be. Everyone's reasons were different, yet all found paradise in the same place.

Entering the world of moneyless living can be pretty scary. But what real adventure isn't? Did humans make their biggest discoveries by staying comfortable? The good news for anyone who wants to explore is that more and more people are walking this path, putting up signposts, laying stepping stones, writing guidebooks. All anyone has to do is to decide that they want to go. That is the most difficult part.

This book is a very rough map of the forest. The money-free life is an adventure; and like any adventure, you should put the map away every now and again and see where the path leads. I recommend, if you are interested in exploring this way of living, that you find your own way. Every one of us is different and we live in different communities. There's no one solution for all; our solutions must be locally driven to meet the needs of people and the environment in which they live. Living without money was how we all lived once, but that was a long time ago.

None of us are teachers; we are all students, learning from each other's experience. I hope you find something in mine. Take what you find useful and stick the rest in the recycling bin of ideas.

USEFUL WEBSITES

BookHopper (www.bookhopper.com)

Book Crossing (www.bookcrossing.com)

Couchsurfing (www.couchsurfing.com)

Carshare (www.carshare.com and www.nationalcarshare.com)

Fergus Drennan (www.wildmanwildfood.com)

Free text messages: (www.cbfsms.com)

Freecycle (www.freecycle.org)

Freegle (www.ilovefreegle.org)

Freelender (www.freelender.org)

Global Freeloaders (www.globalfreeloaders.com)

GROFUN (www.grofun.org.uk)

Gumtree (www.gumtree.com)

Hospitality Club (www.hospitalityclub.org)

LETS (www.letslinkuk.org)

Liftshare (www.liftshare.com)

Money Saving Expert (www.moneysavingexpert.com)

ReaditSwapit (www.readitswapit.co.uk)
Selfsufficientish.com (www.selfsufficientish.com)
Skype (www.skype.com)
Streets Alive (www.streetsalive.net)
SUSTRANS (www.sustrans.org.uk)
Swishing (www.swishing.org)
Swaparamarazzmatazz
 (www.myspace.com/swaparamarazzmatazz)
The Freeconomy Community (www.justfortheloveofit.org)
The Ramblers Association (www.ramblers.org.uk)
Timebank (www.timebanking.org)
Transition Culture (www.transitionculture.org)
UK Freegans (www.freegan.org.uk)
World Wide Opportunities on Organic Farms (WWOOFing)
 (www.wwoof.org)

INDEX

INDEX

INDEX